Energy and Environment Research in China

More information about this series at http://www.springer.com/series/11888

Wenfeng Shangguan · Guchu Zou
Zhi Jiang

Simultaneous Catalytic Removal of Diesel Soot and NO_x

上海交通大學出版社
Shanghai Jiao Tong University Press

Springer

Wenfeng Shangguan
Shanghai Jiao Tong University
Shanghai
China

Zhi Jiang
Shanghai Jiao Tong University
Shanghai
China

Guchu Zou
Shanghai Institute of Ceramics, Chinese Academy of Sciences (SICCAS)
Shanghai
China

ISSN 2197-0238 ISSN 2197-0246 (electronic)
Energy and Environment Research in China
ISBN 978-981-10-7265-9 ISBN 978-981-10-7266-6 (eBook)
https://doi.org/10.1007/978-981-10-7266-6

Jointly published with Shanghai Jiao Tong University Press, Shanghai, China

The print edition is not for sale in China Mainland. Customers from China Mainland please order the print book from: Shanghai Jiao Tong University Press.

Library of Congress Control Number: 2018944332

Printed on acid-free paper

This Springer imprint is published by the registered company Springer Nature Singapore Pte Ltd.
The registered company address is: 152 Beach Road, #21-01/04 Gateway East, Singapore 189721, Singapore

Preface

In recent years, there has been a dramatic increase in the pressure to develop more efficient, environment-friendly engines for heavy and light vehicles. Diesel engines have attracted significant attention due to higher efficiency and lower cost compared with gasoline; and yet, its emission of nitrogen oxides (NO_x) and soot particulates is likely to cause a negative impact on the environment and human health. Urea-SCR has been selected by European and USA motor industry as an emissions control technology system for diesel engines. The system, usually, contains a Diesel Particulate Filter (DPF), diesel oxidation catalyst (DOC), SCR catalysts, and the reductant. Despite its successful application to the heavy-duty vehicle, integration of the SCR within the exhaust is still a challenge, considering the extensive infrastructure for the whole system. Thus, the multifunctional combination of DPF and SCR in a single device is currently receiving considerable attention, as it presents several advantages, such as compact and cost-effective (if we can find suitable catalysts). Simultaneous catalytic removal of soot and NO_x can be regarded as a combined process of PM trapping, soot oxidation, and NO_x reduction by soot. If realized, this should be the most desirable after-treatment of diesel exhausts because it is capable of simultaneously removing both harmful substances by using a compact unit.

This book provides a comprehensive overview of what is called “Simultaneous Catalytic Removal of Diesel Soot and NO_x”; It begins with a discussion of the technology in the framework of the current context, covers the fundamental aspects of the simultaneous catalytic removal process of soot and NO_x (a variety of material, reaction kinetics), and also analyzes the external condition influencing the reaction. It provides a new view to interpret soot abatement material, understand the reaction process, and inspire the scientists to design a new catalyst with moderate redox ability. Some original pictures and illustration are also inclusive.

I expect the book will be of particular interest to and suitable for both engineers and scientists in the field of material chemistry, chemical engineering, and automobile industry.

I appreciate the people without whom the book would have been impossible. My study on the catalysis concerning diesel soot and NO_x began in 1993 when I went to Prof. Kagawa's laboratory at Nagasaki University for my study abroad. I sincerely thank Professor Shuichi Kagawa and Professor Yasutake Teraoka for their help and patience in guiding my doctoral research on the control of pollution from diesel engines, which was of great benefit to my research work later. I also dedicate the book to the memory of Prof. Teraoka who passed away in the pride of his life.

I would like to thank my co-authors, colleagues, and students for their contribution to the subsequent research. I also wish to thank the Shanghai Jiao Tong University Press and Springer Press for their support.

Finally, the authors will be grateful for pointing out any shortcomings in the book.

Shanghai, China Wenfeng Shangguan

Contents

1 Introduction . . . 1
1.1 Characterization of Diesel Emission . . . 1
1.2 Catalytic Abatement of NO_x . . . 4
1.2.1 Direct Decomposition of NO_x . . . 4
1.2.2 NH_3 Selective Catalytic Reduction (NH_3-SCR) . . . 5
1.2.3 Hydrocarbon Selective Catalytic Reduction (HC-SCR) . . . 5
1.3 After Treatment of Diesel Soot . . . 5
1.3.1 Diesel Oxidation Catalyst (DOC) . . . 5
1.3.2 Catalytic Diesel Particulate Filter (CDPF) . . . 6
1.4 Simultaneous Removal of Soot and NO_x . . . 6
References . . . 7

2 Catalytic Materials for Simultaneous NO_x–Soot Removal . . . 9
2.1 Spinel-Type Oxides . . . 9
2.1.1 Structure of Spinel Structure Oxides . . . 9
2.1.2 Reaction Profiles and Performance Evaluation . . . 10
2.1.3 Comparison Between Spinels and Simple Metal Oxides . . . 11
2.1.4 Catalytic Performance of Spinel-Type Oxides . . . 12
2.1.5 Modification of $CuFe_2O_4$ Catalysts by Doping of Alkali Metal Cations and Pt . . . 14
2.1.6 Spinel-Type Copper Chromites . . . 18
2.2 Perovskite-Type Oxides . . . 21
2.2.1 Structure of Perovskite-Type Oxides . . . 21
2.2.2 Reaction Profiles and Performance Evaluation . . . 22
2.2.3 The Catalytic Performance for Various Perovskite-Type Oxides . . . 22
2.2.4 Insight into $La_{1-x}A_xB_yMn_{1-y}O_3$ Material . . . 24

2.3 Hydrotalcites-Derived Oxides . . . 29
2.3.1 Structure of Hydrotalcites-Derived Mixed Oxides . . . 29
2.3.2 The Catalytic Performance and Characterization of Co–Al HT (CAO) . . . 30
2.3.3 Modification and Optimization of MAlO (M = Co, Ni, Cu) . . . 32
2.4 Rare-Earth-Based Complex Oxides . . . 35
2.4.1 Properties of Rare-Earth-Based Oxides . . . 35
2.4.2 Co–La Composite Oxides for Simultaneous Soot–NO_x Removal . . . 36
2.4.3 Co–Ce Composite Oxides for Simultaneous Soot–NO_x Removal . . . 40
2.5 Ordered Mesoporous Oxide Material for Simultaneous Soot–NO_x Removal . . . 47
2.5.1 Catalytic Performance of Mesoporous Co_3O_4 . . . 48
2.5.2 Insight into Mechanism for Co_3O_4 Mesoporous Material . . . 50
References . . . 64

3 Kinetics Study for Simultaneous Removal of Soot and NO_x . . . 71
3.1 Kinetic Characteristics of Simultaneous NO_x–Soot Removal . . . 71
3.1.1 Arrhenius Plots and Compensation Effects . . . 71
3.1.2 Kinetic of Simultaneous NO_x–Soot Removal Over $CuFe_2O_4$ . . . 74
3.1.3 Kinetic of Simultaneous NO_x–Soot Removal Over K-Substituted $CuFe_2O_4$ . . . 82
3.2 Numerical Model of NO_x–Soot Removal . . . 85
3.2.1 Mathematical Model . . . 86
3.2.2 Simulation . . . 87
3.3 Theoretical Study of PM–NO_x Catalytic Reaction . . . 88
3.3.1 Insight into the Interaction Between NO and Soot . . . 88
3.3.2 DFT Calculation for the Adsorption of NO Over $CuFe_2O_4$ . . . 93
References . . . 99

4 Influencing Factors for Simultaneous NO_x–Soot Removal . . . 101
4.1 Contact Conditions . . . 101
4.1.1 Static State . . . 102
4.1.2 Dynamic State . . . 107
4.2 Precious Metal Loading . . . 111
4.3 Surface Area . . . 119
References . . . 123

5 Conclusion and Prospects . . . 127

About the Authors

Wenfeng Shangguan is a professor of Research Center for Combustion and Environment Technology, Shanghai Jiao Tong University (SJTU), China. He received his Ph.D. degree in applied physical chemistry from Nagasaki University, Japan, in 1996. He worked as research fellow during 1996–2000 in Kyushu National Industrial Research Institute, Japan. He was promoted to a professor of SJTU in 2000. His research interests include environmental catalysis, combustion and diesel exhaust emission control, photocatalysis and solar hydrogen production. He has published more than 200 peer-reviewed papers. He is on the list of the most cited Chinese researchers by Elsevier China. He is the holder of more than 30 patents. He is a guest professor at Kyushu University, Japan. He serves as the committee member at Catalysis Society of China, and as the member of editorial board for *Journal of Environmental Science*, *Frontiers in Energy, etc.*

Guchu Zou obtained his Ph.D. degree from Shanghai Jiao Tong University (SJTU) in 2016, is now serving as the research associate in Key Laboratory of Inorganic Coating Materials CAS, Shanghai Institute of Ceramics (SICCAS). He has also worked at Aston University for one year. His research covers a variety of environmental fields, remarkably, environmental catalysis. Simultaneous removal of soot and NO_x via catalytic technology is one of his focuses in his Ph.D. thesis. Several papers addressing the properties of cobalt-based catalysts and their specific roles in the desoot and deNO_x reaction have been recognized and published.

Zhi Jiang is an associate professor of Research Center for Combustion and Environment Technology, Shanghai Jiao Tong University, China. He received the Bachelor and Ph.D. degree in physical chemistry from University of Science and Technology of China. He has ever worked at Hong Kong Polytechnic University, New Jersey Institute of Technology, Purdue University (West Lafayette) and Aston

University (UK). His research interests include solar fuel, metal-based reactive fuel, and environmental catalysis. He is a guest associate professor of Kyushu University, Japan, and young committee member of IEHB for Chinese Society for Environmental Sciences.

Chapter 1
Introduction

Diesel engines have been widely popularized as a power source for vehicles owing to its reliable horsepower and excellent fuel economy. In Europe, especially, they occupy an enormous market share in public transportation and private uses. However, the pollution caused by diesel-engine exhausts has become more and more severe in the last decade, while the exhausts of gasoline-fueled engines operated near stoichiometric air/fuel ratio have been successfully cleaned up by three-way catalytic systems. In China, the emission of diesel particulate matter has become one of the primary sources of $PM_{2.5}$ pollution in cities. The emission from diesel engines is still a teaser, which bothering the scientific researchers and policymakers for years and it is affected by many factors such as the horsepower of vehicles, the composition of diesel, and combustion parameters. The properties of diesel emission are entirely different from that of a gasoline engine, which has considerably been alleviated by the three-way catalytic (TWC) system. The trade-off relations between the two pollutants (PM and NO_x) and oxygen-rich atmosphere disable the TWC, posing a higher requirement for the researchers. Several catalytic techniques have been proposed; they can be classified into two groups, that is, individual removal of the two contaminants and simultaneously PM/ NO_x abatement.

1.1 Characterization of Diesel Emission

Over the last 30 years, various catalytic processes have been involved in the after-treatment module to alleviate the exhaust emissions from diesel vehicles. The latter is dominant pollutants in the atmospheric environment, namely particulate matter (PM) and nitrogen oxides (NO_x). Soot is the major component of PM. Its formation is governed by kinetics and thermodynamics in unsteady combustion, whereas NO_x is produced in an oxygen-rich and high-temperature environment [1].

W. Shangguan et al., *Simultaneous Catalytic Removal of Diesel Soot and NO_x*, Energy and Environment Research in China,
https://doi.org/10.1007/978-981-10-7266-6_1

Soot and NO_x are significant challenges for both human health and atmospheric environment [2], and soot is a dominant source of $PM_{2.5}$ emission [3].

NO_x is a widely known pollutant that causes several environmental nuisance events in the 1950s to 1980s; many countries or institutes issue relevant legislation to put a limit on it. In China, specifically, the overall NO_x emission has been significantly reduced since 2012 (Table 1.1). Tracing to the NO_x exhaust source, the absolute contribution value progressively decays, yet the percentage for the contribution of automobile emission steadily rises. Thus, an increasing emphasis placed on NO_x emission from vehicles is essential and urgent. The control of $PM_{2.5}$ emission has also been ahead of schedule due to its enormous threat to human health. It is estimated that $PM_{2.5}$ is of strong correlation with mortality in the USA [4]. On a global scale, the mortality caused by $PM_{2.5}$ comes up to a peak value in both east China and north India [5]. The mortality can be somehow inhibited via cutting $PM_{2.5}$ emission. In Europe, 11,612 cardiopulmonary deaths and 1901 lung cancer deaths could be prevented annually if long-term exposure to $PM_{2.5}$ levels is reduced to 15 μg/m^3 in each city [6]. Regarding $PM_{2.5}$ contributions, Table 1.2 presents a regular practice. Overall, the $PM_{2.5}$ contribution of automotive is diverse

Table 1.1 NO_x emission amount from various emission sources

Emission sources	2012		2013		2014		2015	
	(million tons)	(%)	(million tons)	(%)	(million tons)	(%)	(million tons)	(%)
Industry	16.581	70.926	15.457	69.398	14.048	67.603	11.809	64.453
Life	0.393	1.681	0.407	1.827	0.451	2.170	0.651	3.553
Vehicles	6.400	27.376	6.405	28.757	6.278	30.212	5.859	31.978
Others	0.004	0.017	0.004	0.018	0.003	0.014	0.003	0.016
Total	23.378	100.000	22.273	100.000	20.780	100.000	18.322	100.000

Data acquired from the China Ministry of Environment Protection

Table 1.2 $PM_{2.5}$ emission sources of eight typical regions in China from 2013 to 2015

Region	External sources (%)	Local emission sources				
		Coal (%)	Automotive (%)	Industry (%)	Fugitive dust (%)	Others (%)
Beijing	28–36	22.4	31.1	18.1	14.3	14.1
Shanghai	16–36	13.5	29.2	28.9	13.4	15
Hangzhou	18–38	18.8	28	22.8	20.4	10
Nanjing	19.6–37.9	27.4	24.6	19.0	14.1	14.9
Guangzhou	n/a	20.6	21.7	11.5	10.4	35.8
Shijiazhuang	23–30	28.5	15	25.2	22.5	8.8
Tianjin	22–34	27	20	17	30	6
Ningbo	n/a	n/a	22	47	11	n/a

Data acquired from the China Ministry of Environment Protection

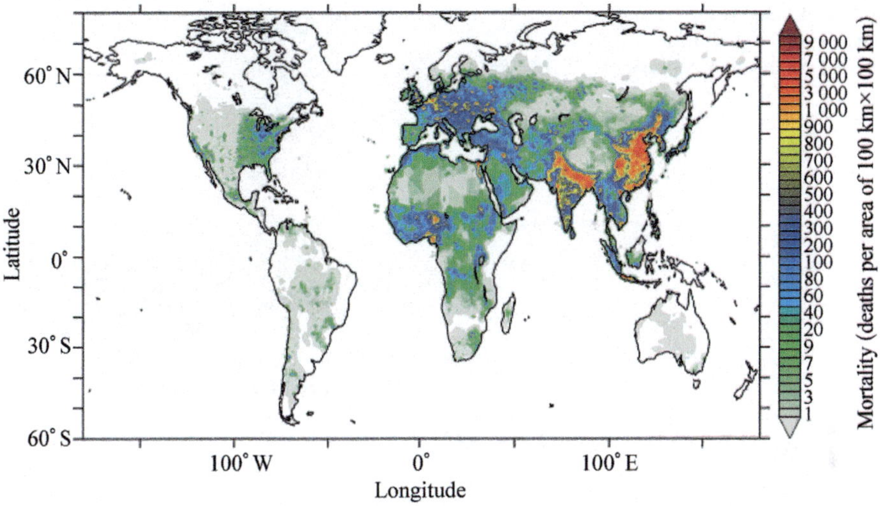

Fig. 1.1 Mortality linked to outdoor air pollution in 2010 [5]. Units of mortality: deaths per area of 100 km × 100 km (color-coded). In the white areas, annual mean $PM_{2.5}$ and O_3 are below the concentration-response thresholds where no excess mortality is expected

in all regions owing to the level of economic development, terrain feature, industrial structure, and energy consumption structure. It can be observed that in population-densified areas, such as Beijing, Shanghai, Hangzhou, and Nanjing, automobile emission is dominant in $PM_{2.5}$ contribution. In most regions, the automotive contribution to $PM_{2.5}$ exhaust is above 20%. On the other hand, regions rely on heavy industry and coal-oriented industry struggles from $PM_{2.5}$ pollution from coal sources. It can be summarized from Table 1.2 that automobile is becoming a non-negligible factor for PM emission. Furthermore, the boom of auto industry also threats the human health and urbanization process. Thus, legislation and technology are required to restrict pollutant emission to follow the pace of this trend (Fig. 1.1).

The relationship between soot and NO for different combustion concepts is shown in Fig. 1.2. Unfortunately, the trade-off relationship between these two pollutants inhibits the use of the three-way catalyst (TWC) on diesel engines; hence, several technical strategies were proposed to deal with the contaminants. These routes can be classified into $deNO_x$ and desoot techniques.

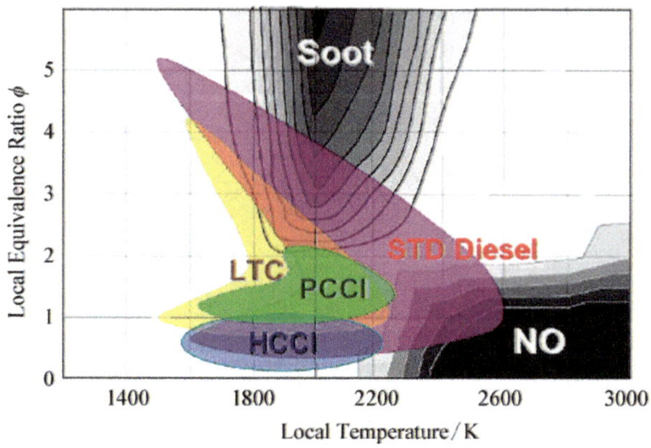

Fig. 1.2 Relationship between soot and NO for different combustion concepts [7]

1.2 Catalytic Abatement of NO_x

Three typical catalytic processes including NO_x decomposition, NH_3-SCR, and HC-SCR are introduced in this section.

1.2.1 Direct Decomposition of NO_x

This approach has been explored for years. Judging by the value of Gibbs free energy, this is a spontaneous reaction. However, the activation energy value for this decomposition process achieves as high as 364 kJ/mol, which indicates that this reaction is unlikely to proceed under mild conditions [8].

$$2NO \rightarrow N_2 + O_2 \quad \Delta G_f^0 = -86\ \text{kJ/mol} \tag{1.1}$$

$$NO_{ads} \rightarrow N_{ads} + O_{ads} \tag{1.2}$$

$$2N_{ads} \rightarrow N_2 \tag{1.3}$$

$$2O_{ads} \rightarrow O_2 \tag{1.4}$$

The direct decomposition of NO on material surfaces usually involves a dissociative chemisorption step, followed by the recombination of dissociated species. Among the above three essential steps for NO_x, the O_2 desorption occurring above 500 °C is the determinant step. In the oxygen-rich atmosphere, O_2 tends to take up the adsorption site and hinders further adsorption of NO. A variety of catalysts including PGM, mixed metal oxides, and zeolites were developed and investigated [9–13]. The Cu-ZSM-5 material is one of the promising candidates owing to its high conversion rate. However, this technical strategy hardly fits the practical operation conditions, for instance, high O_2 content, sulfur poisoning, water vapor, and temperature window of exhaust gas.

1.2.2 NH_3 Selective Catalytic Reduction (NH_3-SCR)

NH_3-SCR was formerly implemented as a deNO_x technology for stationary sources; NO_x is reduced by NH_3 in oxygen-rich surroundings over the V_2O_3/TiO_2 catalyst, and the target product is nitrogen and water. Moreover, sulfur tolerance of TiO_2-based supports is superior to most of the materials; hence, this catalyst is regarded as a potential candidate loaded on heavy-duty diesel vehicles. The storage and transportation are thorny issues for ammonia. In turn, Urea-SCR strategy has been taken into account as a substitute owing to its safety and convenience in China. The shortcoming of Urea-SCR is also evident. The formation of high polymer suppresses the NO_x and NH_3 adsorption on the active site.

$$NH_3 + NO + O_2 \rightarrow N_2 + H_2O \tag{1.5}$$

1.2.3 Hydrocarbon Selective Catalytic Reduction (HC-SCR)

At the end of last century, the selective reduction of NO by hydrocarbon was discovered and proposed by Iwamoto [14]. Massive materials-reductant combinations have been tested to improve the performance of the reaction. Among these candidates, Ag/Al_2O_3-ethanol system is a potential option. Its catalytic activity, the water tolerance, and sulfur tolerance meet the requirements of practical use; furthermore, the reductant is easily acquired [15, 16]. Unfortunately, this concept has not been widely accepted by manufacturers due to the high cost of the post-process system, the storage, and supplement of reducing agent.

$$HC + NO_x \rightarrow N_2 + H_2O + CO_2 \tag{1.6}$$

1.3 After Treatment of Diesel Soot

Two strategies for soot removal are presented in this section.

1.3.1 Diesel Oxidation Catalyst (DOC)

The application of DOC, whose active phase is Pt–Pd-based precious gold metal (PGM), can be traced to 1970s. The DOC catalyzes the following three reactions. It can remove most of the soluble organic fractions (SOF) and reduce the overall PM emission by 25–50%.

$$[SOF] + O_2 \rightarrow CO_2 + H_2O \tag{1.7}$$

$$CO + O_2 \rightarrow CO_2 \tag{1.8}$$

$$[HC] + O_2 \rightarrow CO_2 + H_2O \tag{1.9}$$

Unfortunately, DOC is negligent in limiting soot emission and secondary particles formation. Besides, the oxidation of sulfur into SO_3 at higher temperature leads to the deposition of metal sulfate and in turn poisons the catalyst. Judging by the statement above, neither PM nor NO_x can be purified by DOC module.

1.3.2 Catalytic Diesel Particulate Filter (CDPF)

DPF is a physical way to filter and trap soot via its micropores, mesopore, and macropores; however, it needs to regenerate timely within the temperature window to maintain working efficiency. The constituents supported on the CDPF system are preferentially designed. Thousands of literature have reported competitive catalysts including mixed metal oxides, PGM, and rare earth metal oxides. Most of these studies focused on the mechanism and pathways how soot is consumed. The key to the soot abatement can be summarized into two points, that is, produce sufficient oxidizing agent/intermediates, and further improvement of contact conditions between soot and active sites.

$$PM + O_2 \rightarrow CO_2 \tag{1.10}$$

1.4 Simultaneous Removal of Soot and NO_x

The aforementioned catalytic processes and technologies can be categorized into one-way catalyst. Namely, they can only assist soot gasification or NO_x removal individually. The concept of simultaneous abatement of soot and NO_x was firstly proposed by Yoshida [17]. A redox reaction proceeds between soot and NO_x; the targeted product is CO_2 and N_2, which is harmless to the environment and human health.

$$PM + NO_x \rightarrow CO_2 + N_2 \tag{1.11}$$

Few systematic studies were performed in this field. Shangguan and his group were the first to carry out such laboratory research [18, 19]. These studies are of great significance for not only an academic purpose but also for the practical situation. Firstly, the insight into the reaction provides another view to interpret the catalysts in the reaction. This reaction posed higher requirements to the catalyst.

Specifically, a balance between the oxidability and reducibility is to be weighed and considered. Secondly, the kinetics of the reaction inspires the engineers to design electronic control unit (ECU) program to monitor and adjust atmosphere to accelerate the reaction and suppress the formation of by-products (N_2O). Finally, this reaction is affected by several factors such as contact conditions, flow rate, and reactant gas. These factors are to be carefully evaluated and regarded as the base data of the catalyst. The above three aspects will be detailed demonstrated in the following chapters.

References

1. Z. Mansurov, Soot formation in combustion processes (review). Combust. Explos. Shock Waves **41**, 727–744 (2005)
2. M. Kampa, E. Castanas, Human health effects of air pollution. Environ. Pollut. **151**, 362–367 (2008)
3. M. Zheng, G.R. Cass, J.J. Schauer, E.S. Edgerton, Source apportionment of PM2. 5 in the southeastern United States using solvent-extractable organic compounds as tracers. Environ. Sci. Technol. **36**, 2361–2371 (2002)
4. M. Franklin, A. Zeka, J. Schwartz, Association between PM2. 5 and all-cause and specific-cause mortality in 27 US communities. J. Expo. Sci. Environ. Epidemiol. **17**, 279–287 (2007)
5. J. Lelieveld, J. Evans, M. Fnais, D. Giannadaki, A. Pozzer, The contribution of outdoor air pollution sources to premature mortality on a global scale. Nature **525**, 367–371 (2015)
6. E. Boldo, S. Medina, A. Le Tertre, F. Hurley, H.G. Mücke, F. Ballester, I. Aguilera, Apheis: Health impact assessment of long-term exposure to PM2. 5 in 23 European cities. Eur. J. Epidemiol. **21**, 449–458 (2006)
7. B. Wang, S. Mosbach, S. Schmutzhard, S. Shuai, Y. Huang, M. Kraft, Modelling soot formation from wall films in a gasoline direct injection engine using a detailed population balance model. Appl. Energy **163**, 154–166 (2016)
8. A. Fritz, V. Pitchon, The current state of research on automotive lean NO_x catalysis. Appl. Catal. B **13**, 1–25 (1997)
9. M. Haneda, Y. Kintaichi, N. Bion, H. Hamada, Alkali metal-doped cobalt oxide catalysts for NO decomposition. Appl. Catal. B **46**, 473–482 (2003)
10. Y. Teraoka, T. Harada, S. Kagawa, Reaction mechanism of direct decomposition of nitric oxide over Co-and Mn-based perovskite-type oxides. J. Chem. Soc. Faraday Trans. **94**, 1887–1891 (1998)
11. S. Iwamoto, R. Takahashi, M. Inoue, Direct decomposition of nitric oxide over Ba catalysts supported on CeO_2-based mixed oxides. Appl. Catal. B **70**, 146–150 (2007)
12. R. Gopalakrishnan, P.R. Stafford, J.E. Davidson, W.C. Hecker, C.H. Bartholomew, Selective catalytic reduction of nitric oxide by propane in oxidizing atmosphere over copper-exchanged zeolites. Appl. Catal. B **2**, 165–182 (1993)
13. S. Tsujimoto, X. Wang, T. Masui, N. Imanaka, Direct Decomposition of NO into N_2 and O_2 on C-type Cubic Y_2O_3-ZrO_2 and Y_2O_3-ZrO_2-BaO. Bull. Chem. Soc. Jpn. **84**, 807–811 (2011)
14. M. Iwamoto, Symposium on catalytic technology for removal of nitrogen oxides. Catal. Soc. Jpn. 17–22 (1990)
15. T. Miyadera, Alumina-supported silver catalysts for the selective reduction of nitric oxide with propene and oxygen-containing organic compounds. Appl. Catal. B **2**, 199–205 (1993)

16. S. Sumiya, M. Saito, H. He, Q.-C. Feng, N. Takezawa, K. Yoshida, Reduction of lean NO_x by ethanol over Ag/Al_2O_3 catalysts in the presence of H_2O and SO_2. Catal. Lett. **50**, 87–91 (1998)
17. K. Yoshida, S. Makino, S. Sumiya, G. Muramatsu, R. Helferich, Simultaneous reduction of NO_x and particulate emissions from diesel engine exhaust. SAE Technical Paper (1989)
18. Y. Teraoka, K. Nakano, S. Kagawa, W. Shangguan, Simultaneous removal of nitrogen oxides and diesel soot particulates catalyzed by perovskite-type oxides. Appl. Catal. B **5**, L181–L185 (1995)
19. W. Shangguan, Y. Teraoka, S. Kagawa, Simultaneous catalytic removal of NO_x and diesel soot particulates over ternary AB_2O_4 spinel-type oxides. Appl. Catal. B **8**, 217–227 (1996)

Chapter 2
Catalytic Materials for Simultaneous NO_x–Soot Removal

The catalysts are the major challenge for this scheme. Both soot oxidation and NO_x reduction should be considered and evaluated; it is tough to balance the two aspects mentioned above. Also, the material is required to be thermally stable in oxygen-rich conditions up to high temperature (above 700 °C); in this way, it could survive the thermal load impact. Naturally, oxides are firstly tested due to their thermal stability and activity. The rational design and development of catalysts for this scheme initiate from 1989 by Yoshida [1] followed by the proposal for catalysts of the various type including spinel-type oxides, perovskite-type oxides, hydrotalcite-derived oxides, rare-earth-based oxides, and ordered mesoporous oxides. These materials are displayed in this section.

2.1 Spinel-Type Oxides

2.1.1 Structure of Spinel Structure Oxides

The structure of the spinel-type material (AB_2O_4) is depicted in Fig. 2.1. An octahedral framework of composition BO_2 is derived from NaCl structure by alternate spacing rows of metal ions. Metal ions (A) fill in positions of tetrahedral coordination. The crystallographic unit cell of the spinel structure contains 32 O atoms in approximately cubic close packing (CCP). In another spinel-type material called inverse spinel, the location of A and B ions completely alter. Precisely, the eight tetrahedral positions are occupied by 8 B ions out of 16, while the rest 8 B ions and 8 A ions take up octahedral positions randomly. This type of spinel is formulated $B(AB)O_4$ to distinguish them from those of the first category spinel, AB_2O_4. The spinel structure is somewhat "hollow" since 1/8 of the tetrahedral and 1/2 of the octahedral sites are occupied. The solubility of the component oxides in the spinel phase is also discovered, resulting in the appearance of excess cations; the

W. Shangguan et al., *Simultaneous Catalytic Removal of Diesel Soot and NO_x*, Energy and Environment Research in China,
https://doi.org/10.1007/978-981-10-7266-6_2

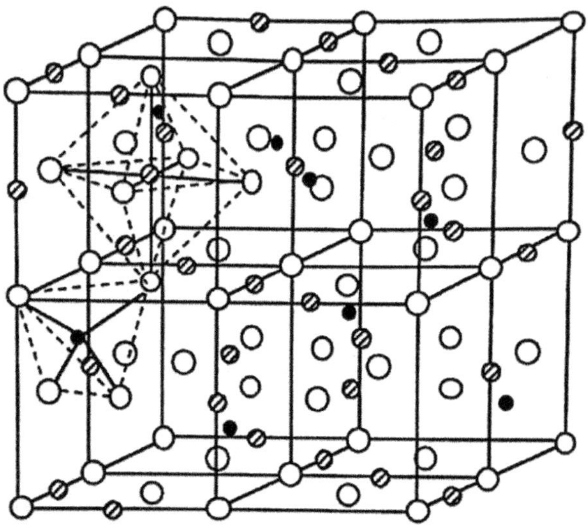

Fig. 2.1 Normal spinel structure

given cations in the spinel structure are substantially substituted by other ions, which provide the possibility to modify its structure further to meet a variety of application, such as catalyst [2–4].

2.1.2 Reaction Profiles and Performance Evaluation

One of the realistic systems for catalytic reduction of PM and PM trap loaded with a catalyst (catalytic PM trap). There are many technical difficulties in this system such as the development of an efficient and thermally stable PM trap, loading of catalytic material on the trap, and the contact between the catalyst and the trapped PM. Judging by the fact that the experimental difficulty of continuous supply of solid soot into a reaction system, the present study was carried out by a temperature-programmed reaction in which a mixture of soot and catalyst was heated in a flow of reactant gas [5].

The profiles for simultaneous soot–NO_x removal over $CuFe_2O_4$ are presented in Fig. 2.2. The formation of CO_2 due to the oxidation of soot and the reduction of NO_x into N_2 and N_2O were detected. It is noteworthy that the three observed products show the similar tendency, indicating soot is consumed by NO_x rather than O_2.

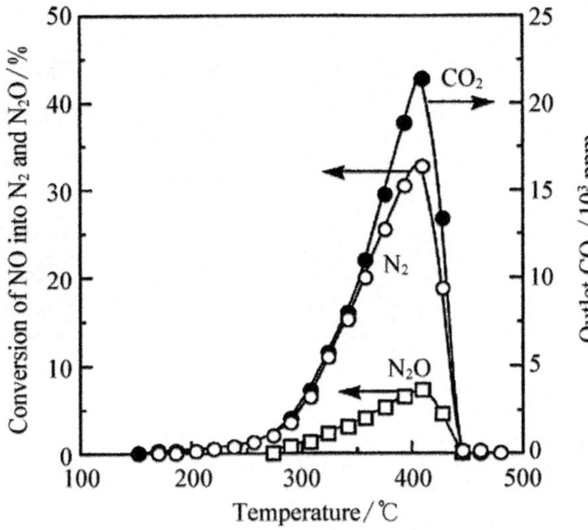

Fig. 2.2 Temperature dependence of NO_x conversion into N_2, N_2O, and CO_2 formed throughout TPO run in the simultaneous removal of NO_x and diesel soot particulate over $CuFe_2O_4$ [5]. Ramp: 1 °C/min; gas composition: NO 0.5%, O_2 5%, balanced with He; flow rate: 20 mL/min

The reaction involved in the TPO process

$$(1-\alpha-\beta)C+(1-\alpha-\beta)O_2 \longrightarrow \alpha N_2+2\beta N_2O+CO_2 \quad (2.1)$$

$$\alpha C+2\alpha NO \longrightarrow \alpha N_2+\alpha CO_2 \quad (2.2)$$

$$\beta C+4\beta NO \longrightarrow 2\beta N_2O+\beta CO_2 \quad (2.3)$$

Overall reaction

$$C+2(\alpha+2\beta)NO+(1-\alpha-\beta)O_2 \longrightarrow \alpha N_2+2\beta N_2O+CO_2 \quad (2.4)$$

T_{ig}: Ignition temperature

$$S_{N_2/C}=V[N_2]/V[CO_2]=\alpha \quad (2.5)$$

$$S_{N_2O}=V[N_2O]/(V[N_2O]+V[N_2])=2\beta/(\alpha+2\beta) \quad (2.6)$$

2.1.3 Comparison Between Spinels and Simple Metal Oxides

A comparison study of simple oxides and spinel structure materials was carried out by Shangguan et al. [5] over Cu–Fe system and Cu–Cr system. T_{ig} value is a parameter to evaluate catalytic performance (Fig. 2.3).

In Cu–Fe system, T_{ig} of $CuFe_2O_4$ and the mechanical mixture (CuO + Fe_2O_3) are close; it is between the two simple oxides. Furthermore, the selectivity to N_2O exhibits no obvious distinction over Cu–Fe samples. These results demonstrate that

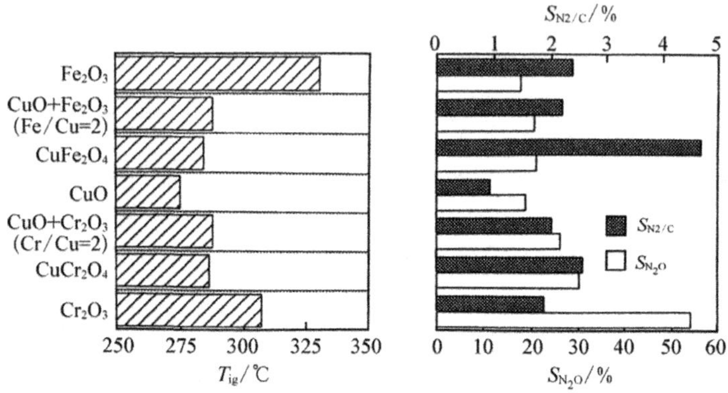

Fig. 2.3 Catalytic performance for the simultaneous NO_x–soot removal of simple oxides, mechanical mixture, and spinel-type oxides over Cu–Fe and Cu–Cr systems

only superimposed effect exists on the activity and the selectivity to N_2O. On the other hand, the prominent high $S_{N_2/C}$ value of $CuFe_2O_4$ implies the synergistic effect by forming the spinel structure in the selectivity of NO_x reduction into N_2. The higher $S_{N_2/C}$ value than mechanical mixture samples presages that atomically mixed state of Cu and Fe ions, such as –Cu–O–Fe– structure, is positive for the promotion of reaction steps including the activation of NO_x and the transfer of activated NO_x to soot surface. The same story applies to the Cu–Cr system, though the synergy in the selectivity to NO_x reduction into N_2 is more pronounced than that in the Cu–Fe system. In contrast, the synergetic action is not observed for the activity but the reduction process over the Cu–Cr system. These above results show the advantage of forming of spinel structure metal oxides in the promotion of $S_{N_2/C}$.

2.1.4 Catalytic Performance of Spinel-Type Oxides

The catalytic performance depends significantly on constituent metal cations. It is evident that Co- and Mn-containing spinels with a lower T_{ig} and $V[N_2]$ are more active but less selective. Among the ternary spinel oxides studied, $CuFe_2O_4$ showed prominently high $V[N_2]$, $S_{N_2/C}$ and low $S_{N_2/C}$ so that $CuFe_2O_4$ with the moderate T_{ig} and the maximum selectivity to the N_2 formation is the proper catalyst for the simultaneous NO_x–soot removal (Fig. 2.4).

The effect of metal cations and their coordination in spinel oxides will be briefly discussed. Divalent Co ions are in the octahedral coordination in the inverse $CoFe_2O_4$ spinel, while being in the tetrahedral coordination in normal $CoCr_2O_4$ and $CoMn_2O_4$. As stated above, these Co-containing spinels exhibit low and comparable selectivity to nitrogen formation. In addition, it is apparent from the comparison between CuM_2O_4 and CoM_2O_4 (M = Cr, Fe) that the replacement of Co with Cu brings about an enhancement of the selectivity to N_2 formation irrespective

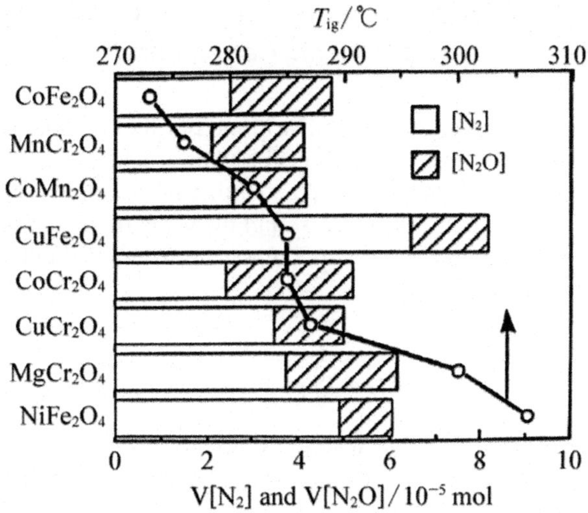

Fig. 2.4 Ignition temperature of soot (T_{ig}) and the amount of N_2 and N_2O formed over the spinel-type material [5]

of tetrahedral (M = Cr) or octahedral (M = Fe) coordination. These results indicate that the sort of metal cations is crucial, whereas the coordination state of metal cations is of less importance.

Catalytic performance of spinel-type oxides is shown in Table 2.1.

Table 2.1 Catalytic performance of spinel-type oxides in the simultaneous NO_x–soot removal reaction

Catalyst	T_{ig} ① (°C)	$V[N_2]$ ② (10^{-5} mol)	Selectivity (%)	
			$S_{N2/C}$ ③	S_{N2O} ④
$CuCr_2O_4$	287	3.51	2.6	29.9
$MgCr_2O_4$	300	3.73	2.7	39.4
$CoCr_2O_4$	285	2.43	1.8	53.4
$MnCr_2O_4$	276	2.11	1.5	49.3
$CoMn_2O_4$	282	2.57	1.9	38.7
$NiFe_2O_4$	306	4.90	3.6	19.1
$CoFe_2O_4$	273	2.53	1.8	46.5
$CuFe_2O_4$	285	6.46	4.7	20.8

①Ignition temperature of soot
②Total amount of N_2 in the reaction
③Selectivity to NO into N_2
④Selectivity to N_2O

2.1.5 *Modification of $CuFe_2O_4$ Catalysts by Doping of Alkali Metal Cations and Pt*

In Sect. 2.1.1, the possibility to modify the structure of spinel oxides has been well illustrated. Various modifiers or additives have been utilized to improve the activity and selectivity of the catalysts further, although the specific understanding chemical effect of additives is not straightforward [6, 7]. These promoters bring about the development of the multi-component materials. In this section, several practices are presented.

2.1.5.1 Effect of the Type for Metal Cations Doping

The doped material was prepared by the citric acid-aided process. The catalytic performance depends significantly on the sort of dopant cations (Fig. 2.5). As for alkali metal doping, cesium enhances the activity (decreasing T_{ig}) with the selectivity to the N_2 formation ($V[N_2]$) being almost intact; potassium promotes both the activity and selectivity; sodium improves the activity but deteriorates the selectivity, and lithium causes the deterioration of both the activity and selectivity. The activity of alkali metal-doped oxides increases with atomic number as $Li < Na < K \approx Cs$, being in accordance with the reported order for the NO–C reaction over the alkali metal–carbon systems, $Na < K < Cs$ [8].

Given the promotion of both the activity and selectivity, potassium is the most effective dopant to $CuFe_2O_4$, and the substitution of potassium for Cu or Fe resulted in almost the same promotion effect manifested in Fig. 2.5. It is not surprising since potassium is a popular promoter of Cu-based soot oxidation catalysts [9, 10].

It was reported that V_2O_5 was an active catalyst for the soot oxidation [10], and the Cu–V mixed oxide system was used quite often as the soot oxidation catalysts [11, 12]. Accordingly, the effect of the doping of vanadium to $CuFe_2O_4$ was investigated. As shown in Fig. 2.5, however, the $Cu_{0.9}V_{0.1}FeO_4$ catalyst is inferior

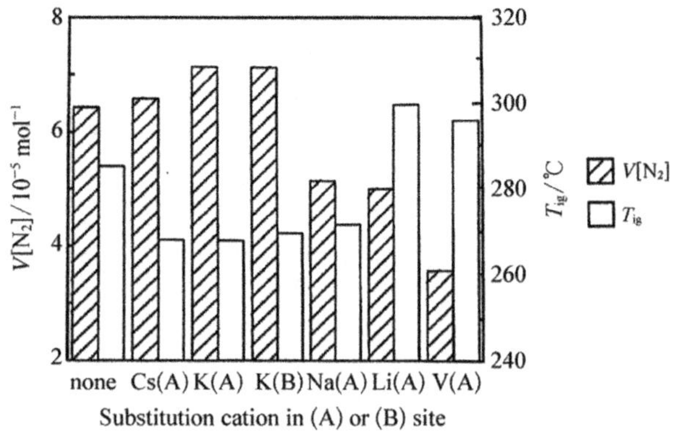

Fig. 2.5 Effect of cation substitution for A-site or B-site of $CuFe_2O_4$ on the ignition temperature of soot (T_{ig}) and the amount of N_2 formed ($V[N_2]$) in the simultaneous NO_x–soot removal None: unsubstituted $CuFe_2O_4$

to $CuFe_2O_4$ in both activity and selectivity, suggesting that vanadium doped in the Cu–Fe spinel-type oxide is adverse to the catalytic performance for the simultaneous NO_x–soot removal.

2.1.5.2 Effect of Potassium Doping Amount

The effect of the doping amount of potassium was investigated for $Cu_{1-x}K_xFe_2O_4$ (Fig. 2.6). The optimized T_{ig} and the maximum $V[N_2]$ are attained simultaneously at $x = 0.05$, and the further doping of potassium caused sharp and progressive decreases in both the activity and the selectivity to the N_2 formation. These results uncover that the optimum doping amount (x) in $Cu_{1-x}K_xFe_2O_4$ is 0.05. The results of impregnated 5 mol% $K/CuFe_2O_4$ catalyst are also shown in Table 2.2. A similar promotion effect of potassium was also observed with the impregnated $K/CuFe_2O_4$ catalyst, though it is inferior to pre-doped $Cu_{0.95}K_{0.05}Fe_2O_4$ in both the activity and selectivity.

Regarding the formation of N_2O, the selectivity to N_2O (S_{N2O}) over $x = 0.05$ (18.6%) is slightly smaller than but comparable to that over $x = 0$ (20.8%). It can thus be confirmed that potassium shows almost no effect on the suppression of the N_2O formation. The effect of S_A will be discussed below.

2.1.5.3 Impregnation with K and Pt

K-impregnated catalysts are as follows: Pulverized $CuFe_2O_4$, α-Al_2O_3, and CuO were immersed in a solution dissolved with CH_3COOK so as for the loading of K to be 5 mol%. Pt-impregnated catalysts were acquired via the following process: Pulverized $CuFe_2O_4$ and $Cu_{0.95}K_{0.05}Fe_2O_4$ were added in an aqueous solution of H_2PtCl_4 in a proportion of 0.5 wt% Pt. The calcination temperature was fixed at 700 °C.

$V[N_2]$ values are revealed against T_{ig} in Fig. 2.7 to provide further interpretation for the effect of the doping of potassium and platinum. The comparison between

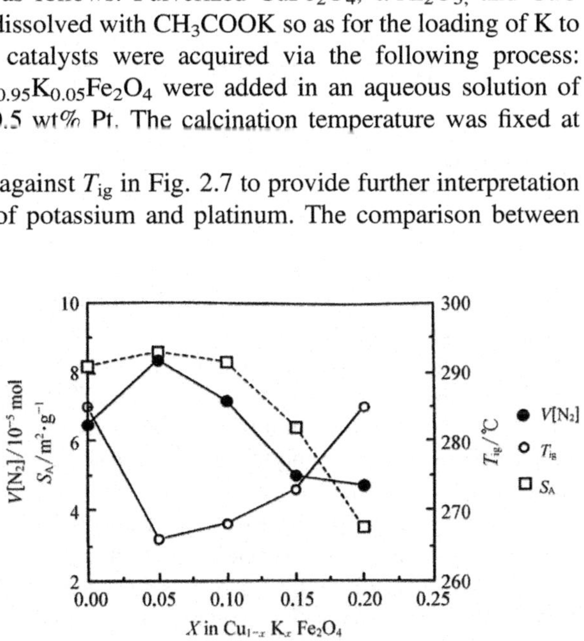

Fig. 2.6 Effect of substitution of K for Cu in $Cu_{1-x}K_xFe_2O_4$ on the T_{ig}, $V[N_2]$ and specific surface area (S_A)

Table 2.2 Catalytic performance of K-, Pt-doped oxides for the simultaneous NO_x–soot removal

	Catalysts	T_{ig} ① (°C)	$V[N_2]$ ② (10^{-5} mol)	$S_{N2/C}$ ③ (%)
1	$CuFe_2O_4$	285	6.46	4.7
2	$Cu_{0.95}K_{0.05}Fe_2O_4$	266	8.34	6.1
3	$K(Ac)/CuFe_2O_4$	270	6.09	4.4
4	$K(Cl)/CuFe_2O_4$	260	3.80	2.8
5	$Pt/CuFe_2O_4$	292	3.50	2.5
6	$Pt/Cu_{0.95}K_{0.05}Fe_2O_4$	258	3.07	2.2
7	α-Al_2O_3	409	6.50	4.7
8	$K(Ac)/\alpha$-Al_2O_3	262	2.79	2.0
9	Pt/γ-Al_2O_3	265	2.08	1.5

①Ignition temperature of soot
②Total amount of N_2 in the reaction
③Rate of the N_2 formation to the rate of soot consumption

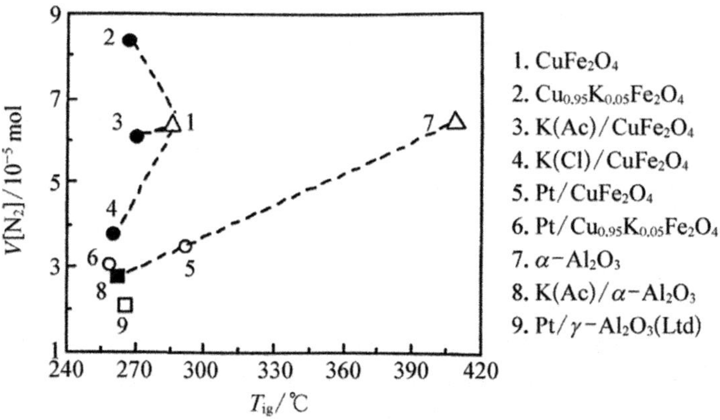

Fig. 2.7 Relation between the amount of $V[N_2]$ and T_{ig} for $CuFe_2O_4$ and α-Al_2O_3 with and without K- and Pt-doping [13]. Broken lines show a pair of K-free (open) and K-doped (closed) oxides

α-Al_2O_3 and K/α-Al_2O_3 shows that the doping of potassium is capable of the promotion of the activity (decreasing T_{ig}). It is a universal consensus in catalytic chemistry that platinum is an active catalyst while potassium is not. The comparable catalytic performance between K/α-Al_2O_3 and Pt/α-Al_2O_3 demonstrates the specific role of potassium in the simultaneous NO_x–soot removal reaction as confirmed in other carbon-involved catalytic reactions such as NO–C reaction, oxidation, and gasification of carbon.

As stated above, pre-doped $Cu_{0.95}K_{0.05}Fe_2O_4$ and impregnated $K/CuFe_2O_4$ catalysts show high activity and selectivity. The impregnation of platinum on $CuFe_2O_4$ and $Cu_{0.95}K_{0.05}Fe_2O_4$ results in a decrease in the selectivity to the N_2 formation ($V[N_2]$) with the activity (T_{ig}) remaining unchanged. This finding indicates that platinum preferentially accelerates the side reaction of the oxidation of soot by oxygen, and therefore, it is not an active ingredient for the simultaneous removal of NO_x and soot. Since the K/CuO catalyst also performs poorly in the selectivity, it can be concluded that the high catalytic performance, especially high selectivity, is realized as the proper amount of potassium is doped in the Cu–Fe spinel catalyst.

2.1.5.4 Characterization of K-Promoted Catalysts

All the K-doped catalysts showed XRD patterns similar to the K-absence $CuFe_2O_4$. Since XRD peaks were weak and broad, however, it is impossible to judge whether or not potassium is dissolved in the crystal lattice. The specific surface area of $Cu_{1-x}K_xFe_2O_4$ barely varies at $x \leq 0.1$ and decreased above that point. The deterioration of catalytic performance at higher substitution level is not ascribed to the decrease in specific surface area, but to the surface concentration of K. The ionic radius of K^+ (0.151 nm) is larger than that of Cu^{2+} (0.071 nm); hence, there is not much the chance that the complete substitution of K^+ for Cu^{2+} occurs; it can be deduced that the overall substitution level is quite limited. The excess potassium may concentrate or segregate on the surface of spinels. The catalytic performance for the simultaneous NO_x–soot removal reaction was barely affected by the specific surface area of catalysts since the reaction takes place at the so-called "triple contact point" where the solid catalyst, the solid soot, and the gaseous reactants meet together [14]. The surface concentration of potassium is more dominant for the x-dependence of catalytic performance than that of the specific surface area.

Surface states of $Cu_{1-x}K_xFe_2O_4$ (x = 0–0.2) and $K/CuFe_2O_4$ were analyzed by XPS. Binding energy (B. E.) of $2p_{3/2}$ lines of Cu, Fe, or K was almost the same for all the samples; 933.7–934.0 eV (Cu), 710.6–710.9 eV (Fe), and 292.6–292.8 eV (K). Judging from the B. E. values, copper, iron, and potassium are in divalent [15], trivalent [16] and monovalent states, respectively, and their oxidation states (B. E. values) remain unchanged before and after the catalytic reaction. Overall compositions of some oxides were measured by atomic absorption spectrometry (AAS), and they are consistent with the stoichiometric ratio. On the other hand, the surface composition is quite different from the overall composition. As shown in Fig. 2.8, the surface composition of potassium is quite higher than the overall (weighted-in) composition and increased with an increase in the doping amount. It is noted that surface composition of potassium for pre-doped $Cu_{0.95}K_{0.05}Fe_2O_4$ and impregnated 5 mol% $K/CuFe_2O_4$ is relatively close. This phenomenon indicates that potassium is not incorporated into the crystal lattice of the Cu–Fe spinel oxide and segregates on the surface of spinel grains. The high activity and selectivity are attained only in the combination of potassium and the Cu–Fe spinel at the proper doping amount, which suggests that the cooperative or synergetic action of potassium and the Cu–Fe spinel is necessary, and this situation may be realized at an intermediate and appropriate doping amount (x = 0.05) or surface composition of potassium.

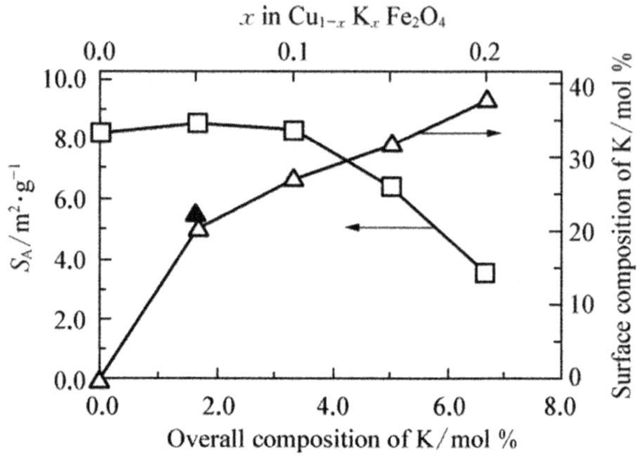

Fig. 2.8 S_A and surface content as a function of potassium doping amount in $Cu_{1-x}K_xFe_2O_4$ [13]. Closed triangle: 5% mol/K/$CuFe_2O_4$

Excessive doping of potassium results in an increase in the surface concentration of potassium which causes the coverage for the active sites of the Cu–Fe spinel oxide.

2.1.6 Spinel-Type Copper Chromites

2.1.6.1 Catalytic Performance for Simultaneous NO_x–Soot Removal Over $CuCr_2O_4$

The reaction profiles are shown in Fig. 2.9 which are similar to those of above-mentioned spinel-type $CuFe_2O_4$. The parallelism that is the nearly the same E_a, between N_2 and CO_2 formation rates suggests that the rate-determining step of N_2 and CO_2 formation is the same. The selectivity to NO_x reduction is only 2.9%

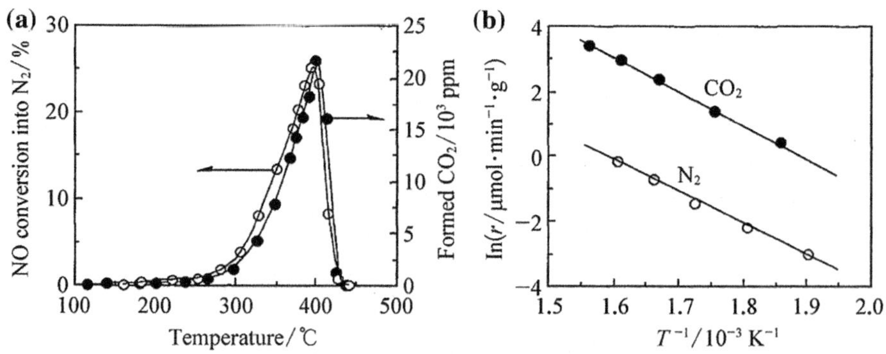

Fig. 2.9 Simultaneous removal of NO_x and diesel soot particulate over $CuCr_2O_4$ [17]. **a** Temperature dependence of NO_x conversion into N_2 and the outlet concentration of CO_2; **b** Arrhenius-type plots of N_2 and CO_2 formation rates

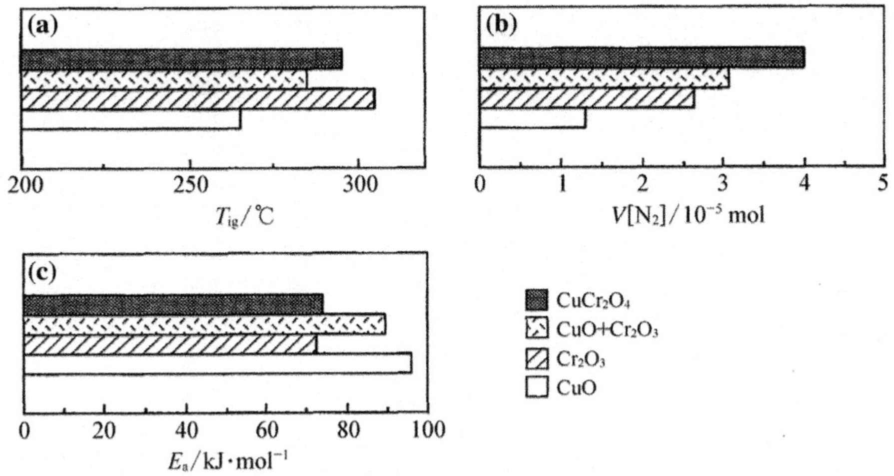

Fig. 2.10 Comparison of ignition temperature of soot (**a**), the total amount of N_2 formed (**b**), and apparent activation energy for N_2 formation (**c**) of Cu–Cr oxides [17]

for $CuCr_2O_4$ indicating that the oxidation of the soot by O_2 predominates over that by NO_x in the present reaction system.

Catalytic activity for the simultaneous removal reaction of simple oxides (CuO, Cr_2O_3), the mechanical mixture (CuO + Cr_2O_3, Cr/Cu = 2 in molar ratio) and spinel-type oxide ($CuCr_2O_4$) was measured, and three parameters, T_{ig}, $V[N_2]$, and E_a for N_2 formation are derived from the reaction results (Fig. 2.10). T_{ig} reaches the lowest value with CuO, followed by the CuO + Cr_2O_3 mixture, spinel-type $CuCr_2O_4$, and Cr_2O_3. These results indicate that regarding the catalytic activity, the spinel-type oxide and the mechanical mixture containing Cu and Cr lie between active CuO and less active Cr_2O_3 on the whole. Regarding the selectivity to NO_x reduction, on the other hand, catalysts with Cu and Cr contents are superior to CuO and Cr_2O_3 indicating that the coexistence of Cu and Cr ions is beneficial to the selectivity.

2.1.6.2 Effect of Substitution on the Simultaneous NO_x–Soot Removal Over $CuCr_2O_4$

The effect of partial replacement in the present $CuCr_2O_4$ spinel was also investigated. Catalytic parameters T_{ig}, $V[N_2]$ and the selectivity to NO_x reduction, of $CuCr_2O_4$-based oxides with various substituents, are listed in Table 2.3, and the T_{ig} versus $V[N_2]$ relation of catalysts except for less active Sn-substituted oxide is shown in Fig. 2.11. The catalytic performance of these oxides varies dramatically depending on both A- and B-site cations as well as the substitution level; a spinel-type oxide is generally expressed as AB_2O_4, and therefore A- and B-sites mean Cu and Cr sites, respectively, in the present case. Roughly speaking, catalysts

Table 2.3 Catalytic activities of substituted Cu–Cr spinel-type oxides for the simultaneous removal of NO_x and soot particulate [17]

	Catalyst①	T_{ig}② (°C)	$V[N_2]$③ (10^{-5} mol)	α④ (%)
1	$CuCr_2O_4$	295	4.00	2.9
2	$Cu_{0.9}Li_{0.1}Cr_2O_4$	285	2.11	1.5
3	$CuCr_{1.9}Li_{0.1}O_4$	260	2.36	1.7
4	$Cu_{0.8}Na_{0.2}Cr_2O_4$	280	1.30	0.9
5	$CuCr_{1.8}Na_{0.2}O_4$	268	1.04	0.8
6	$Cu_{0.9}K_{0.1}Cr_2O_4$	282	2.90	2.1
7	$Cu_{0.8}K_{0.2}Cr_2O_4$	280	2.50	1.8
8	$CuCr_{1.95}K_{0.05}O_4$	284	3.77	2.7
9	$CuCr_{1.8}K_{0.2}O_4$	255	2.56	1.9
10	$CuCr_{1.8}Sn_{0.2}O_4$	370	2.93	2.1
11	$CuCr_{1.9}Co_{0.1}O_4$	290	4.57	3.3
12	$Cu_{0.9}Fe_{0.1}Cr_2O_4$	300	4.21	3.1
13	$CuCr_{1.9}Mn_{0.1}O_4$	285	4.60	3.3

①Prepared by the citric acid-aided process at 700 °C
②Ignition temperature of soot
③The total amount of N_2 formed during the reaction
④Selectivity to NO_x reduction

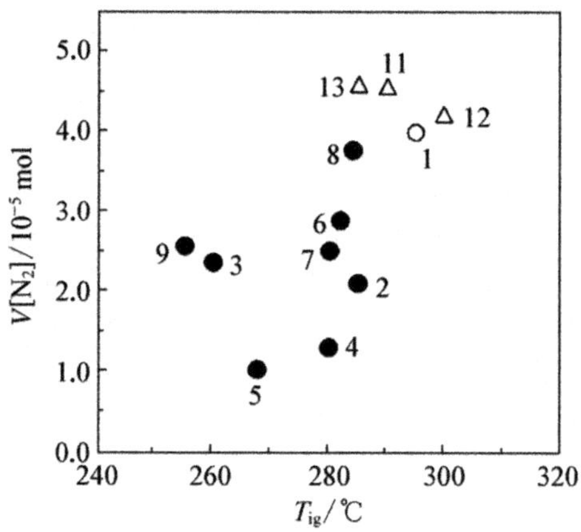

Fig. 2.11 Relation between ignition temperature of soot (T_{ig}) and the total amount of N_2 formed [17] ($V[N_2]$) of substituted Cu–Cr spinel-type oxides prepared by the CIT process at 700 °C, refer to Table 2.3 for the listing of catalysts

substituted by transition metal cations are comparable to unsubstituted $CuCr_2O_4$ with respective to activity and selectivity. The substitution of alkali metal cations, on the other hand, causes a deactivation (a decrease in T_{ig}), and the effect of B-site substitution is more prominent than that of A-site substitution. However, the incorporation of alkali metal ions results in a decrease in selectivity, though effects of the sort and the composition of substituted alkali metal cations are not straightforward. The effect of the substitution of alkali metal ions in $CuCr_2O_4$ observed in this study was entirely different from that found in perovskite-type oxides [18]. These differences are not due to the crystal structure but to constituent metal cations, since the promotion effect in spinel-type $CuFe_2O_4$ is similar to that in perovskite-type oxides.

The ignition temperature is also determined by the reaction gas as $NO + O_2$ (295 °C) < O_2 (325 °C) < NO (520 °C). The minimum T_{ig}, corresponding to the highest activity, in $NO + O_2$ gas further suggests such a mechanism that the reaction is accelerated by the cooperation of NO and O_2, for example, the implication of NO_2, especially in lower temperature region.

2.2 Perovskite-Type Oxides

2.2.1 Structure of Perovskite-Type Oxides

The typical perovskite structure (ABO_3) illustrated in Fig. 2.12 is cubic, in which A is surrounded by 12 O atoms and B by 6 O atoms. Comparatively, few compounds meet these criteria while the lattice structures of many compounds are slightly distorted variants with lower symmetry. There exists a series of perovskite-related oxides expressed by $(ABO_3)_nAO$. They are characterized by alternate stacking of n unit cell and one unit cell thick of rock salt-type AO.

The initial application of perovskite oxides for the catalysis of the desired reaction is electrocatalysis. Since then, the material attracted extensive attention as redox catalysts for automobile exhaust treatment [19]. The perovskite containing lanthanide in position A and a transition metal in position B has been discovered to be a promising catalyst for a variety of reactions such as CO/hydrocarbon/oxygenated compounds oxidation, NO_x reduction [20–23].

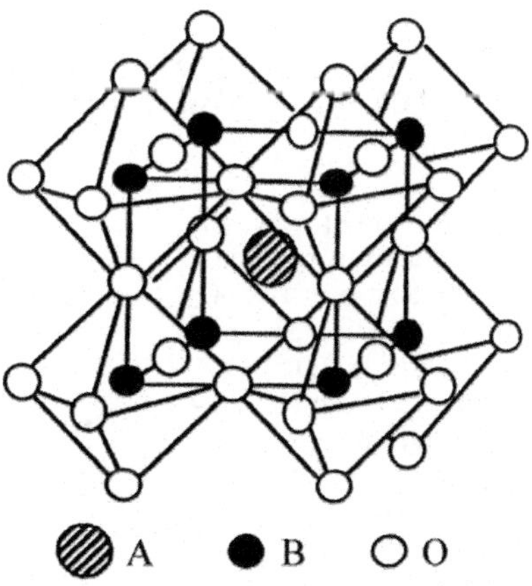

Fig. 2.12 Crystal structures of perovskite-type (ABO_3)

2.2.2 Reaction Profiles and Performance Evaluation

A typical reaction profiles are presented in Fig. 2.13 over $LaCoO_3$. The formation of CO_2 due to the consumption of soot and the reduction of NO_x into N_2 are observed at the same temperature range; moreover, the tendency of the two products is similar. This indicates the occurrence of simultaneous removal of NO_x and soot. T_{ig}, $V[N_2]$, and $S_{N2/C}$ are adopted to evaluate the catalytic performance, which has been stated in Sect. 2.1.2.

2.2.3 The Catalytic Performance for Various Perovskite-Type Oxides

Catalytic performance for some typical oxides is summarized in Table 2.4 and Fig. 2.14. Naturally, a quantified catalyst should be the one with low T_{ig} and massive $V[N_2]$, namely it is supposed to be located at the lower left corner in Fig. 2.14. It is observed that the selectivity to NO_x reduction of Pt/Al_2O_3 and other simple transitional oxides are at low level and that perovskite-type oxides are superior to simple oxides on NO_x reduction. The catalytic performance of perovskite-type oxides strongly depends on both A-site and B-site cations. It is essential to understand the effect of both A-site and B-site cations further.

In either series of K- or Sr-substituted oxides, the trade-off relation that $V[N_2]$ tends to decay with the increase of T_{ig} is observed. The catalytic activity for the gas-phase oxidation reaction of Co- and Mn-based perovskite-type oxides is known to be higher than that of Fe-based perovskites [24] and judging from T_{ig}, and this tendency for the catalytic performance of $NO–O_2$–soot reaction still preserved.

The predominant promotion effect with the improvement of $V[N_2]$ and T_{ig} via the modification of A-site substitute from strontium to potassium was uncovered. In

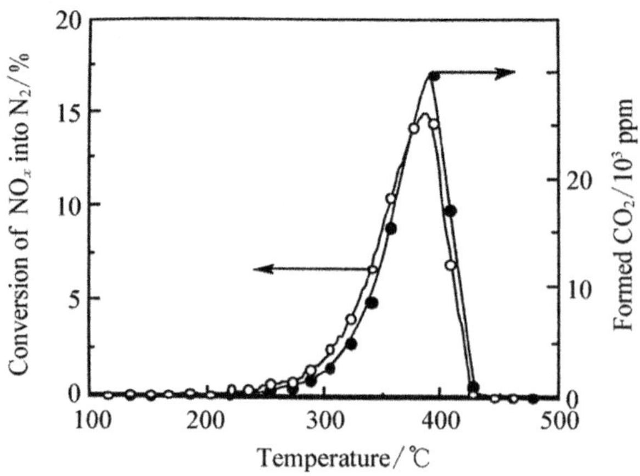

Fig. 2.13 TPO profiles for the simultaneous removal of NO_x and soot over $LaCoO_3$. Ramp: 1 °C/min; gas composition: NO 0.5%, O_2 5% balanced with He; flow rate: 20 mL/min

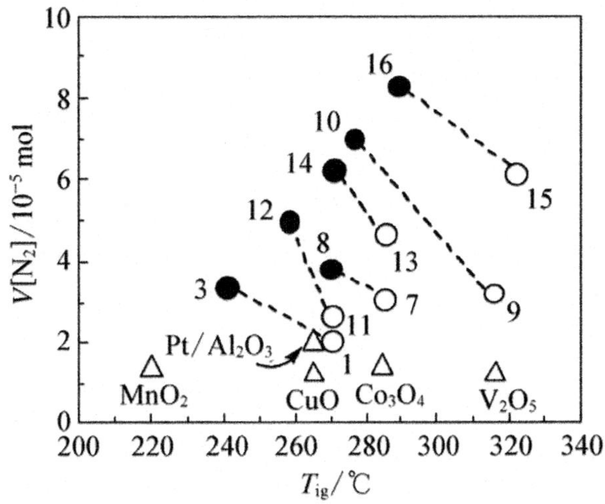

Fig. 2.14 Relation between $V[N_2]$ and T_{ig}, refer to Table 2.3 for the list of catalysts [14]. Broken line: a pair of Sr-substituted (open circle) and K-substituted (closed circle) oxides with the same B-site cations

Table 2.4 Catalytic performance of perovskite-type oxides for the simultaneous NO_x–soot removal [14]

	Catalyst	S_a① (m^2 g)	T_{ig}② (°C)	$V[N_2]$③ (10^{-5} mol)	$S_{N2/C}$④ (%)
1	$La_{0.8}Sr_{0.2}CoO_3$	5.6	270	2.05	1.5
2	$La_{0.95}K_{0.05}CoO_3$	1.3	261	3.18	2.3
3	$La_{0.9}K_{0.1}CoO_3$	0.6	240	3.39	2.5
4	$La_{0.8}K_{0.2}CoO_3$	0.6	278	2.46	1.8
5	$La_{0.9}Li_{0.1}CoO_3$	1.6	262	2.75	2.0
6	$La_{0.9}Sr_{0.1}CoO_3$	0.8	267	3.87	1.8
7	$La_{0.4}Sr_{0.6}MnO_3$	4.0	285	3.04	2.2
8	$La_{0.9}K_{0.1}MnO_3$	2.9	269	3.83	2.8
9	$La_{0.8}Sr_{0.2}FeO_3$	6.7	315	3.21	2.3
10	$La_{0.9}K_{0.1}FeO_3$	3.5	276	6.95	5.1
11	$La_{0.4}Sr_{0.6}Mn_{0.8}Fe_{0.2}O_3$	11.5	270	2.66	1.9
12	$La_{0.9}K_{0.1}Mn_{0.8}Fe_{0.2}O_3$	5.6	258	4.95	3.5
13	$La_{0.4}Sr_{0.6}Mn_{0.8}Cu_{0.2}O_3$	5.7	285	4.64	3.4
14	$La_{0.9}K_{0.1}Mn_{0.8}Cu_{0.2}O_3$	2.4	270	6.2	4.5
15	$La_{0.8}Sr_{0.2}Cu_{0.7}V_{0.3}O_x$ (mixture)	1.4	321	6.05	4.4
16	$La_{0.9}K_{0.1}Cu_{0.7}V_{0.3}O_x$ (mixture)	1.5	288	8.25	6.0

①S_a-BET surface area
②Ignition temperature
③N_2 formation throughout the reaction
④Selectivity to NO_x reduction

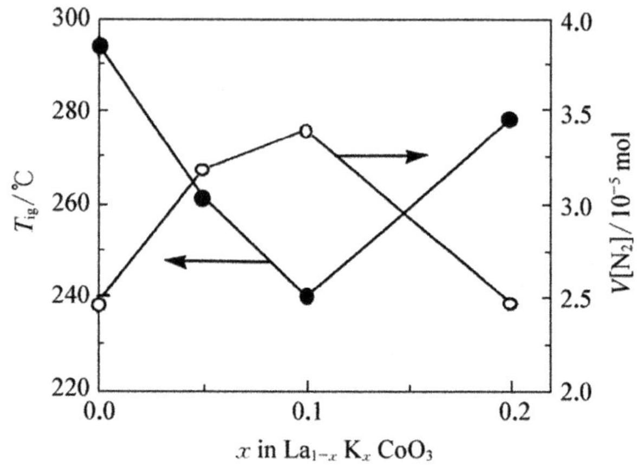

Fig. 2.15 Effect of the substitution of potassium for lanthanum in $La_{1-x}K_xCoO_3$ on T_{ig} and $V[N_2]$

each group of the materials with the same B-site cations which are linked with a broken line in Fig. 2.14, K-substituted oxides exhibit a promotional performance than Sr-substituted oxides do. As presented from the profiles on cobalt-based perovskite-type catalysts (Table 2.4), similar improvement for the performance is also acquired over other alkaline metals (Li, Cs)-substituted samples. Among all the alkaline metals, potassium is the optimized candidate for the doping of perovskite-type oxides. The effect of substitution level of potassium on T_{ig} and $V[N_2]$ over lanthanum-based materials is depicted in Fig. 2.15. The optimum point is fixed at $x = 0.1$. These results reveal that the catalytic performance rest with the incorporation level of potassium. Above all, $La_{0.9}K_{0.1}CoO_3$ with the minimum T_{ig}, while $La_{0.9}K_{0.1}Cu_{0.7}V_{0.3}O_x$ with the maximum $V[N_2]$ is the most selective one.

2.2.4 Insight into $La_{1-x}A_xB_yMn_{1-y}O_3$ Material

The simultaneous soot–NO_x removal is aiming at achieving a low ignition temperature (T_{ig}) and high NO_x conversion to N_2 ($V[N_2]/V[CO_2]$) over a certain material. La–K–Mn–O perovskite-type catalyst was first proposed and investigated by Teraoka et al. in the field of simultaneous soot–NO_x abatement [18]. It was evidenced that this perovskite material is a potential candidate for this reaction.

2.2.4.1 Effect of Cation Substitution

The A-site and B-site substitution over ABO_3 perovskite-type material were performed over $LaMnO_3$ (L–M–O) materials. The effect of cerium dopant on the catalytic activity is presented in Fig. 2.16. The ignition temperatures for $x = 0.2$ and $x = 0.4$ samples are approximately the same; this value is slightly lower than that of

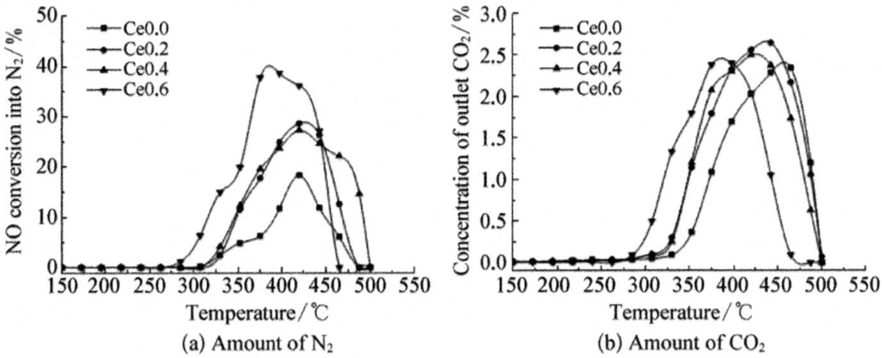

Fig. 2.16 Effect of dopant (Ce) on the catalytic performance of the catalysts [29]

$LaMnO_3$. As the substitution amount further elevates (x = 0.6), the T_{ig} lowers from 330 to 285 °C accompanied by the maximum N_2 formation among all the candidates. Thus, the optimized x value is 0.6 for cerium. Unlike Sect. 2.2.3, a comparison between rare-earth element (Ce) cations and alkali metal ions (K) incorporation into $LaMnO_3$ is made (Fig. 2.17). Apparently, potassium substitution is superior to that of cerium in both catalytic activity and N_2 formation. The divergence electronegativity between K and Ce will result in the valence changes for Mn at B-site. XPS surveys provide additional information. The intensity of Mn 2p increases sharply, which indicates that the surface content of Mn increases. According to the NIST XPS database, the binding energies for the Mn $2p_{3/2}$ peak of Mn_2O_3 and MnO_2 are 641.6 and 642.2 eV, respectively [25–27]. The asymmetry of Mn $2p_{3/2}$ will appear with the coexistence of Mn^{3+} and Mn^{4+}, and this phenomenon will be more predominant as the Mn^{4+}/Mn^{3+} ratio increases [28]. XPS survey illustrates the coexistence of Mn^{4+} and Mn^{3+} on the surface.

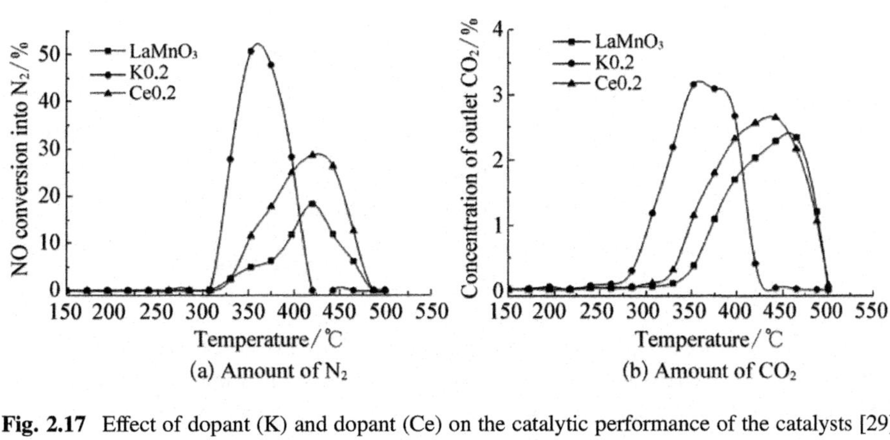

Fig. 2.17 Effect of dopant (K) and dopant (Ce) on the catalytic performance of the catalysts [29]

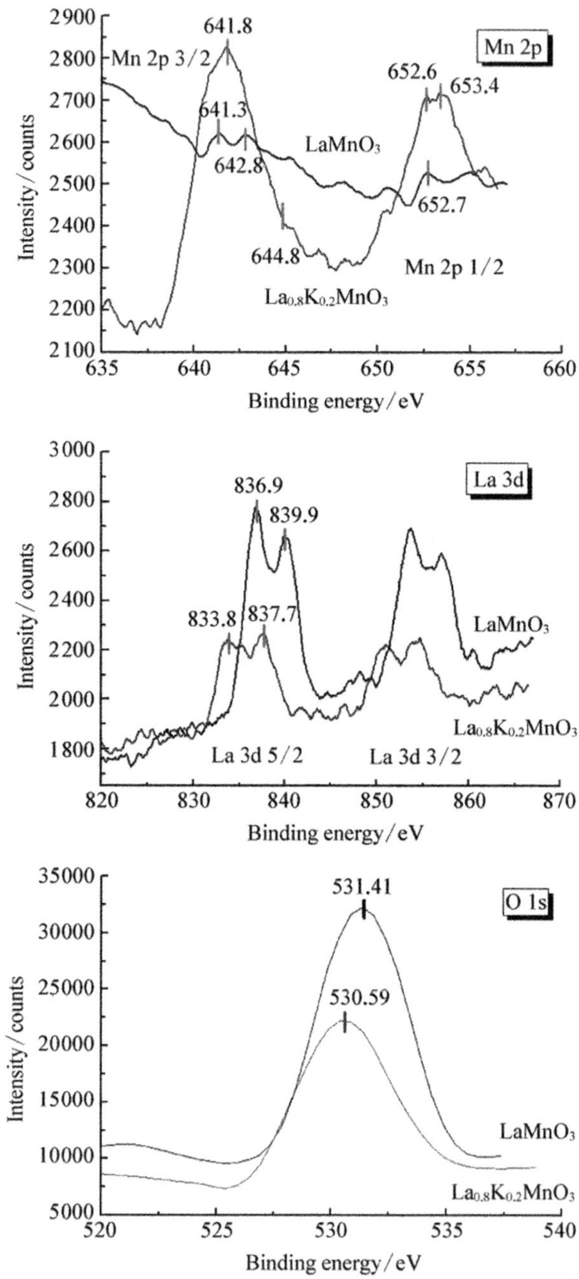

Fig. 2.18 Mn 2p, La 3D and O 1s spectra for K-substituted $LaMnO_3$ and $LaMnO_3$ [29]

This type of substitution results in more crystal defects and then decreases the diffusion resistance derived from the penetrations of gaseous molecules through the capillary among crystal lattices. Thus, the effect of A-site substitution consequently promotes contacts between the gasses and the soot. As a result, it accelerates the complete catalytic reaction.

As stated in previous reports, the optimized substitution level for potassium over $La_{1-x}K_xMnO_3$ is $x = 0.2$ ($La_{0.8}K_{0.2}MnO_3$) [30]. Based on this material, a study on B-site substitution was performed [31]. Figure 2.19 (XRD patterns) shows that there is a small diffraction peak at about 13° as the composite is in the absence of Cu. This peak disappears with the introduction of Cu, which indicates that some Cu promotes the formation of a perfect crystal. Copper is a metal with a low melting point, which can promote the dissolution of different elements in the catalysts during the preparation process. Figure 2.20 also reveals the effects of the partial substitution of Cu for Mn at the B-site. The introduction of copper into $La_{0.8}K_{0.2}MnO_3$ leads to the slight deterioration of ignition temperature of soot along with the uprush of maximum yield of NO to N_2. The maximum yield of NO to N_2 achieves 66.0% (Fig. 2.16 and Table 2.5) as the substitution amount of Cu is 0.05. With the increase of substitution level, the NO removal rate decreases, yet it is still higher than that obtained without Cu substitution.

These results are mainly attributable to the catalytic performances of different cations, such as Cu^{2+} and Mn^{3+}. The incorporation of Cu causes some of the Mn^{3+} become Mn^{4+} to maintain the neutrality of the molecule. XPS survey provides specific clues for this deduction. Regarding La 3d spectra, there is little distinction between $La_{0.8}K_{0.2}MnO_3$- and Cu-substituted samples, which demonstrates the exposure of La^{3+} on the surface. According to NIST database, the binding energy of Mn $2p_{3/2}$ for α-Mn_2O_3 (641.9 eV) and β-MnO_2 (642.2 eV) is approaching. The coexistence of the two compounds intensifies the asymmetry of Mn $2p_{3/2}$ peaks. The asymmetry further aggravates the more Mn^{4+} formation. On the basis of this fact, the Cu introduction leads to the transformation of Mn^{3+} into Mn^{4+} confirmed by Fig. 2.21b.

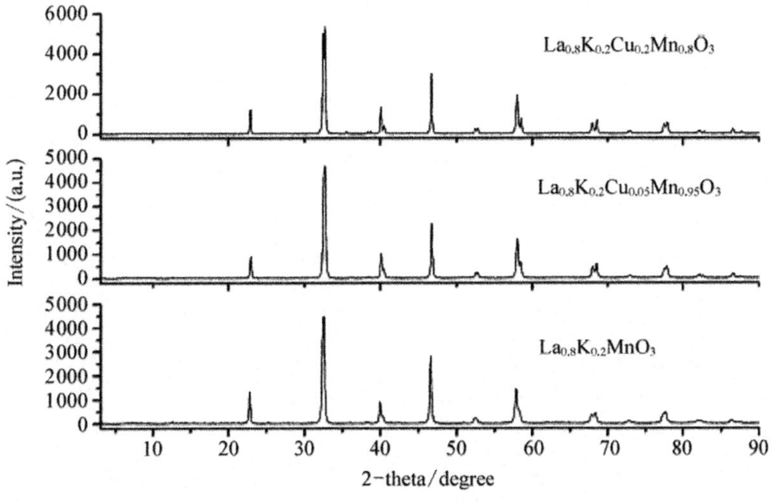

Fig. 2.19 XRD patterns for Cu-substituted $La_{0.8}K_{0.2}Cu_yMn_{1-y}O_3$ [33]

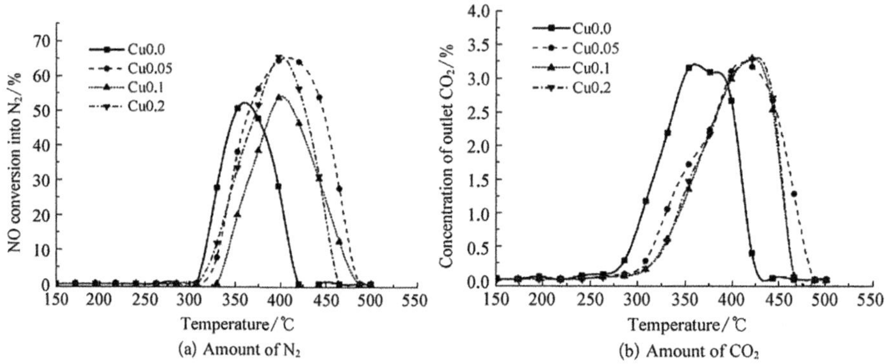

Fig. 2.20 Effect of Cu amount in $La_{0.8}K_{0.2}Cu_yMn_{1-y}O_3$ on the catalytic performance of simultaneous soot–NO_x removal [34]

Table 2.5 Summarized XPS results of Fig. 2.18 [33]

Catalyst	Binding energy (eV)		Surface content (%)				
	La $3d_{5/2}$	Mn $2p_{3/2}$	La	K	Cu	Mn	O
$LaMnO_3$	836.9	641.3	68.56	n/a	n/a	14.09	17.35
$La_{0.8}K_{0.2}MnO_3$	833.8	641.8	63.50	2.46	n/a	23.05	10.99
$La_{0.8}K_{0.2}Cu_{0.05}Mn_{0.95}O_3$	834.4	641.9	55.56	2.00	6.21	23.45	12.78
$La_{0.8}K_{0.2}Cu_{0.2}Mn_{0.8}O_3$	834.5	641.8	49.87	1.54	17.15	18.59	12.85

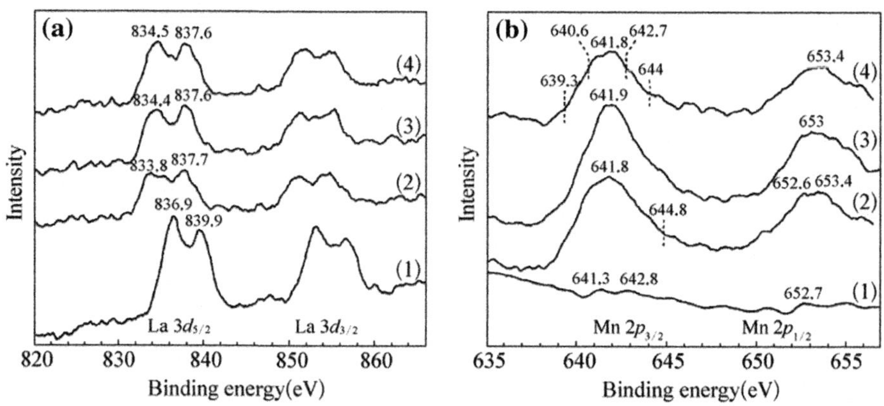

Fig. 2.21 XPS profiles of different catalysts [33]. **a** La 3d; **b** Mn 2p; (1) $LaMnO_3$; (2) $La_{0.8}K_{0.2}MnO_3$; (3) $La_{0.8}K_{0.2}Cu_{0.05}Mn_{0.95}O_3$; (4) $La_{0.8}K_{0.2}Cu_{0.2}Mn_{0.8}O_3$

Table 2.6 Catalytic performance of the catalysts [34]

Catalysts	Ignition temperature (°C)	Maximum NO conversion into N_2 (%)	Selectivity S [N_2] (%)
$LaMnO_3$	330	18.4	0.76
$La_{0.8}K_{0.2}MnO_3$	275	52.5	1.46
$La_{0.8}K_{0.2}Cu_{0.05}Mn_{0.95}O_3$	290	66.0	2.92
$La_{0.8}K_{0.2}Cu_{0.1}Mn_{0.9}O_3$	310	53.8	1.88
$La_{0.8}K_{0.2}Cu_{0.2}Mn_{0.8}O_3$	310	65.5	2.27

The catalytic performances for soot combustion of different cations are diverse (Table 2.6). Cu doping might exert an adverse effect on soot ignition temperature. On the other hand, this substitution also results in defects formation in the crystal, which favors the redox reaction. Furthermore, Cu doping modifies the physical structure of the catalysts. Both the synergistic effect of two different ions at the B-site are favorable for the improvement of catalytic performance, according to the catalyst design of perovskites [32], proper incorporation of Cu content play a decisive role on the catalytic performance of $La_{0.8}K_{0.2}MnO_3$.

2.3 Hydrotalcites-Derived Oxides

2.3.1 *Structure of Hydrotalcites-Derived Mixed Oxides*

LDHs can be viewed as derived from hydroxides of divalent cations with the brucite layer structure $[AdBAdB]_n$, by oxidation or cation replacement in the metal layers (d), so as to give them an excess positive electric charge, and intercalation of extra anion layers (Z) between the hydroxide layers (A, B) to neutralize that charge, resulting in the structure $[AcBZAcB]_n$. LDHs can be formed with a wide variety of anions in the intercalated layers (Z), such as Cl^-, Br^-, NO^{3-}, CO_3^{2-}, SO_4^{2-}, and SeO_4^{2-}. This structure is unusual in solid state chemistry since many materials with similar structure (such as montmorillonite and other clay minerals) have negatively charged main metal layers (c) and positive ions in the intercalated layers (Z) [35].

Hydrotalcite (HT) and hydrotalcite-like compounds, a typical form of LDH structure, are widely applied in catalysis and adsorption due to their changeable compositions and unique structural properties. These materials can be chemically expressed by the formula $[M(II)_{1-x}M(III)_x(OH)_2]^{x+}[A(n\text{-})_{n/x}\cdot mH_2O]^{x-}$, where M(II) represents any divalent metal cation, M(III) any trivalent metal cation, and A an anion [36]. At higher temperatures, hydrotalcite is transformed into mixed metal oxides, which are potential catalysts or catalyst supports owing to their large surface areas, fundamental properties, high metal dispersions, and stability against sintering. In the past few years, the interests in the application of HT-derived mixed oxides as environmental catalysts have significantly increased. Calcined

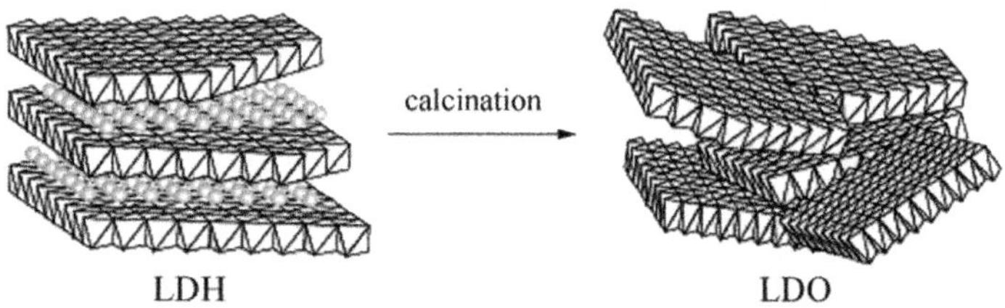

Fig. 2.22 Transformation of layer double hydroxide (LDH) into layer double oxides (LDO)

hydrotalcite has been found to be active and selective catalysts of DeNO$_x$ with ammonia [37], as well as catalytic decomposition of NO and N_2O [38, 39]. Such mixed oxides have also been employed as catalysts or catalyst support in redox reactions by incorporating transition metals, with redox properties, in the HT structure [40]. Their redox performances are strongly linked with the metal species, contents, and calcined temperatures. Copper-, cobalt-, or nickel-based oxides derived from HT are known to be active catalysts for oxidation reactions [41] (Fig. 2.22).

2.3.2 The Catalytic Performance and Characterization of Co–Al HT (CAO)

The major component of calcined CAO is the mixture of Co_3O_4 and $CoAl_2O_4$, which is confirmed by XRD patterns (Fig. 2.23). Figure 2.24 shows the catalytic performances for simultaneous NO_x–soot removal over a variety of CAO catalysts. Both the activity of soot oxidation and the selectivity to the N_2 formation of CAO-800 catalysts are higher than that of CAO-500 catalysts. It may be related to the enhancement of the reducibility caused by higher calcination temperature as mentioned below [42, 43]. Among the two series of the catalysts, the 5CAO-800 sample exhibits the best activity with T_{ig} = 261 °C. It may be linked with a smaller crystallite size of 5CA-HT precursors which results in more uniform particle size of CAO catalysts processed by higher temperature. The uniform particle size of catalysts improves the contacts conditions between soot and catalysts. The oxidation activity is promoted subsequently. Regarding the selectivity to the N_2 formation, the CAO-800 catalysts are also better than CAO-500. Also, there is a decreasing trend for the selectivity to the N_2 formation with growing Co/Al ratio for the two series of catalysts. The 2CAO-800 sample shows the best selectivity to the N_2 formation with $S_{N_2/C}$ = 4.09%, which indicates that the soot oxidation by O_2 is kinetically favored in comparison with NO_x–soot reaction [14, 44]. Furthermore, the selectivity to N_2O over CAO-500 catalysts is 7.3–18.2 and 7.6–23.1% for CAO-800, respectively. As a result, 4CAO-800 is a promising catalyst with

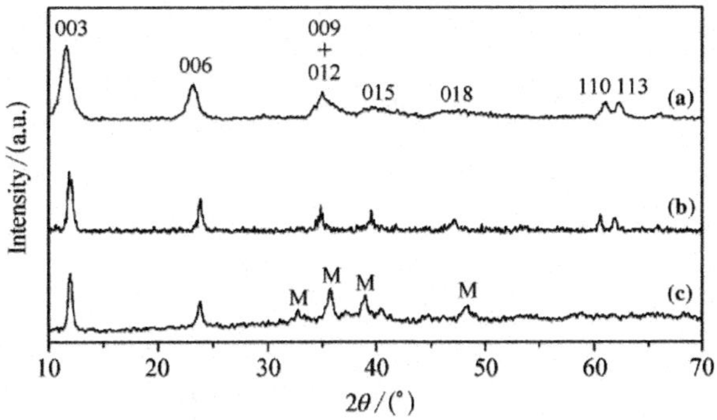

Fig. 2.23 XRD patterns of M(II)–Al HT samples [46]. **a** Ni–Al HT; **b** Co–Al HT; **c** Cu–Al HT. The peaks marked "M" denote malachite $Cu(OH)_2$ phases

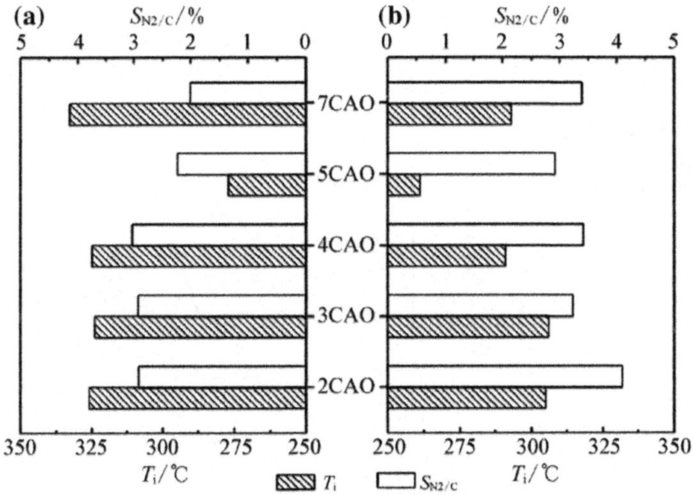

Fig. 2.24 Catalytic performances for the simultaneous NO_x–soot removal over CAO catalysts [46]. **a** CAO-500 samples and **b** CAO-800 samples. The experimental conditions are same as that described in Fig. 2.14

medium activity (T_{ig}= 290 °C) and high selectivity to N_2 formation ($S_{N2/C}$ = 3.5%) as well as low selectivity to N_2O ($S[N_2O]$ = 15.1%) (Figs. 2.24 and 2.25).

Although the specific surface area of CAO-800 samples is quite small, these materials are more active and selective than CAO-500 samples (Table 2.7 and Fig. 2.24), which implies that the catalytic performance is almost independent of the surface area. The simultaneous removal reaction of NO_x and soot supposedly takes place at the so-called triple contact point where the solid catalyst, the solid soot, and the gaseous reactants (NO_x and O_2) meet [5, 14, 18]. On the other hand, features in the temperature range of NO_x–soot reaction (250–500 °C) for CAO catalysts are in accordance with the lower temperature peak of H_2–TPR profiles corresponding to the reduction of Co_3O_4 species (Fig. 2.26). Furthermore, this reaction temperature range is close to that of reference Co_3O_4 spinel under the same

Table 2.7 Composition and textual properties of the samples [46]

Sample	Co/Al molar ratio	Average crystallite size (nm)	Calcined 500 °C			Calcined 800 °C		
			S_{BET} ($m^2\ g^{-1}$)	V_P ($cm^3\ g^{-1}$)	D_P (nm)	S_{BET} ($m^2\ g^{-1}$)	V_P ($cm^3\ g^{-1}$)	D_P (nm)
2CA-HT	2	23.0	114	0.573	20.1	47.0	0.163	13.9
3CA-HT	3	12.6	83	0.413	19.9	30.9	0.160	20.7
4CA-HT	4	9.3	72.2	0.315	17.5	30.9	0.231	29.8
5CA-HT	5	4.0	79.9	0.351	17.6	35.2	0.305	34.7
7CA-HT	7	7.9	64.9	0.385	23.7	18.5	0.211	45.7

The average crystallite size of Co–Al–HT calculated from *d*(003) and *d*(006) planes using Debye–Scherrer equation. S_{BET}—BET specific surface area; V_P—total pore volume; D_P—average pore diameter

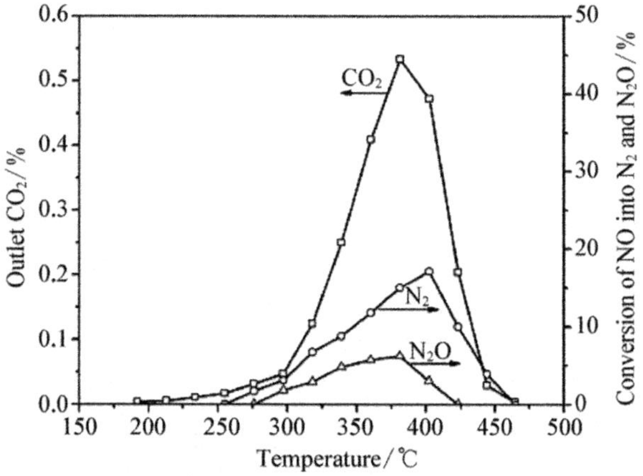

Fig. 2.25 TPO results of the simultaneous NO_x–soot removal over 5CAO-800 [46] Gas composition: NO 0.25 vol.%, O_2 5 vol.%, balanced with He; flow rate: 80 cm^3/min

conditions with T_{ig} = 310 °C and $S_{N2/C}$ = 1.66%. All these facts suggest that the active species in the CAO catalysts are likely to originate from Co_3O_4 which are responsible for the supply of oxygen by a redox reaction. Then there exists a redox-type mechanism for soot oxidation, as was reported by Miro and co-workers who studied Co, K supported catalysts [45]. Compared with the stoichiometric AB_2O_4-type spinel oxides, CAO-800 are medium in activity but more selective to the N_2 formation.

2.3.3 *Modification and Optimization of MAlO (M = Co, Ni, Cu)*

The materials (MAlO) were synthesized via co-precipitation of an aqueous solution of suitable metal nitrates (the molar ratio of M/Al = 3) with an aqueous solution of 2 M NaOH and 1 M Na_2CO_3. The hydrotalcite (HT) structure was confirmed by all the samples. However, there is a significant distinction for copper-based samples in separating out impurity phase, namely $Cu(OH)_2$. The catalytic performances are compared by analyzing the TPO profiles (Fig. 2.27a). It is viewed in Fig. 2.25 that the broadened curves during TPO process in the reaction gas of O_2/He. The non-catalytic soot oxidation initiates at about 530 °C. For potassium-absence samples, all of the ignition temperatures (T_{ig}) for MAlO (M = Ni, Co, and Cu) are about 320 °C and the combustion completes at about 600 C. However, for the potassium-promoted solids, both T_{ig} and T_p shift to lower temperatures. For example, the T_{ig} and T_p of Co-based catalysts decrease to 269 and 350 °C from 316

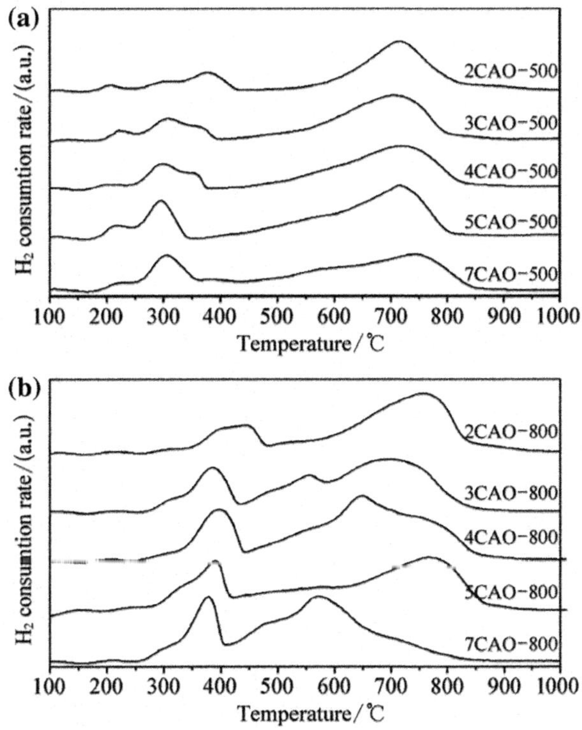

Fig. 2.26 H_2–TPR patterns of CAO samples [46]. **a** CAO-500 samples; **b** CAO-800 samples

Table 2.8 Characteristics of the porous structure of the MAlO and K/MAlO [47]

Sample	S_{BET} ($m^2\ g^{-1}$)	V_P ($cm^3\ g^{-1}$)	D_P (nm)
NiAlO	88.6	0.265	12
CoAlO	35.0	0.104	11.9
CuAlO	17.8	0.052	11.5
K/NiAlO	57.9	0.124	8.5
K/CoAlO	33.9	0.068	8.1
K/CuAlO	17.0	0.049	11.6

S_{BET}—BET specific surface area; V_P—total pore volume; D_P—average pore diameter

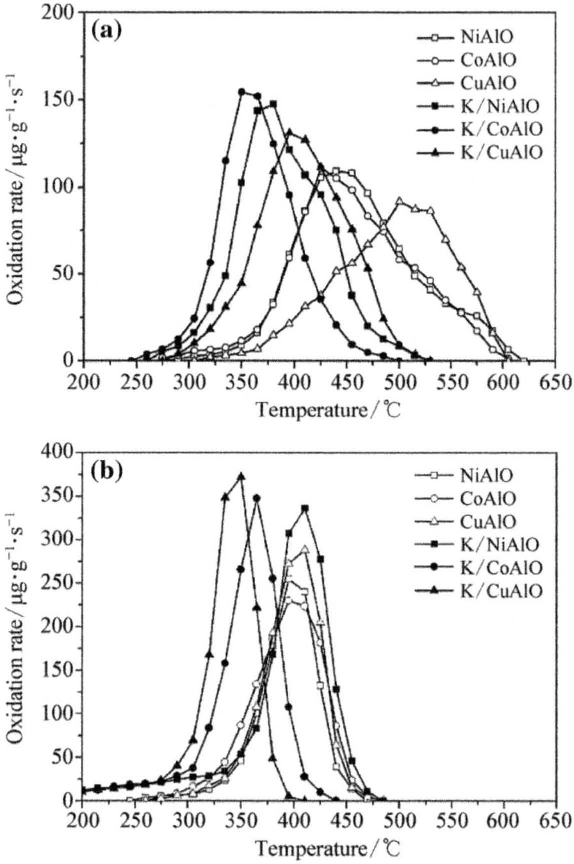

Fig. 2.27 TPO results of soot catalyzed oxidation [47]. Gas composition: **a** 5 vol.% O_2/He, **b** 0.25 vol.% NO + 5 vol.% O_2/He; flow rate: 80 cm^3/min

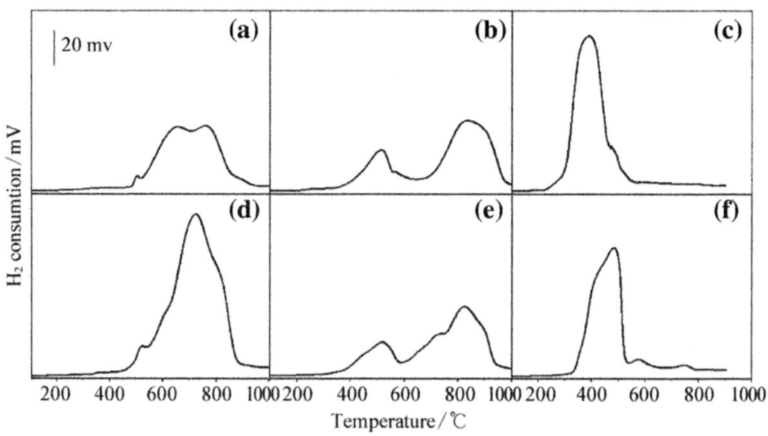

Fig. 2.28 H_2–TPR curves of samples [47]. **a** NiAlO; **b** CoAlO; **c** CuAlO; **d** K/NiAlO; **e** K/CoAlO; **f** K/CuAlO

to 430 °C, respectively, due to the promoting effect of potassium. The promotion effect on soot oxidation is contributed to the increase of surface mobility by loading potassium, improving the contact between the solid catalyst and solid soot [48] (Table 2.8 and Fig. 2.28).

In the presence of NO + O_2 (Fig. 2.27b), the TPO profiles become narrower and sharper compared to the previous one. K–CuAlO sample starts soot combustion at 251 °C and completes this process at about 400 °C in the presence of NO + O_2. In soot–NO–O_2 system, the importance of NO_2 is well known [49]. Even at low temperature, NO can be readily oxidized to NO_2. It has been approved that the oxidizing ability of NO_2 is stronger than NO and O_2. The maximum oxidation rate for soot is considerably increased to about 300 (μg g^{-1} s^{-1}) in the presence of NO, while it is about 150 (μg g^{-1} s^{-1}) in the absence NO. FTIR results evidence the adsorption of NO on the catalyst surface. It seems that the formed nitrate and/or nitrite species in soot–NO–O_2 reactions result in the positive influence on the activity. It has also been reported that the nitrate and/or nitrite are strong oxidizers and contribute to the oxidation of soot [50]. In this study, N_2 and/or N_2O are detected in the TPO process of all catalysts, which indicated that NO_x species participate in the soot oxidation reaction. The simultaneous removal of soot–NO_x removal can be achieved via accelerating soot combustion at the cost of sacrifice $S_{N_2/C}$ value. However, the $V[N_2]$ can still maintain at a high level.

2.4 Rare-Earth-Based Complex Oxides

2.4.1 Properties of Rare-Earth-Based Oxides

17 rare-earth elements, composed of 15 lanthanide elements, scandium (Sc) and Yttrium (Y) are of unique electron structure. The electron fills in 5d spin–orbit rare-earth metal including La, Ce, Lu, and Gd. The 5d orbit is especially crucial for catalysis process such as redox reaction since it is an active platform for electron transfer. Thus, La and Ce are widely applied in the simultaneous soot–NO_x removal in forms of pure oxide, composite oxides, and solid solution oxides. CeO_2 has been widely employed as an automobile catalyst in the three-way catalyst (TWC) system mainly with resort to its unique properties, specifically, the ability to store oxygen, thermal stability and ability to disperse noble metal in its surface. Among the advantages mentioned above, the oxygen storage capacity is an essential attribute to the application of TWC. CeO_2 is capable of storing oxygen at lean burn conditions and can release oxygen when exposed to reducing agents. This above process involves the mutual transformation between Ce^{3+} and Ce^{4+} [51, 52]. On the other hand, La exists in forms of perovskite-type oxides, and spinel-typed oxides which are presented in Sects. 2.1 and 2.2. $LaCoO_3$, a typical composite oxide, have been widely reported in simultaneous soot–NO_x removals [53, 54]. The two instances of rare-earth and transitional metal composite oxides will be introduced below.

2.4.2 Co–La Composite Oxides for Simultaneous Soot–NO_x Removal

Co–La composite oxides are synthesized via CIT method. The cobalt acetate and lanthanum nitrate are used as precursors, and the final calcination temperature is 700 °C. The composite oxides are denoted as $Co_{1-x}La_x$ where x represents the molar ratio of La.

The reaction profiles of $Co_{0.93}La_{0.07}$ are presented in Fig. 2.29. CO_2, N_2, and N_2O are detected. It is noteworthy that even at low temperature (100 °C) a small fraction of NO_x is reduced to N_2. The curve trends are analog with Fig. 2.2, which demonstrates that NO_x instead of O_2 reacts with soot. According to Table 2.9, a proper amount of La introduction leads to the improved T_{10} and N_2 formation (temperature when 10% soot is consumed) over Co–La composite oxides. The optimized x value is 0.07 where T_{10} value reaches 286 °C and N_2 yield at 340 °C (2 $V[N_2]/V[NO]$ * 100%) is 25.2%. It can also be inferred that $LaCoO_3$ is the active phase in Co–La mixed oxides since the T_{10} yield and N_2 yield are close to the optimized value. Furthermore, the activity of $Co_{0.93}La_{0.07}$ is superior to $Co_{0.93}La_{0.07}$–pm sample, which suggests the interaction between two major phases of Co–La mixed oxides.

XRD patterns of all the catalysts are displayed in Fig. 2.30. Two separate phases spinel-type Co_3O_4 and perovskite-type $LaCoO_3$ are detected in the $Co_{0.93}La_{0.07}$ catalysts. The positive shifts of $LaCoO_3$ peaks indicated a lattice shrink over $Co_{0.93}La_{0.07}$ samples. This phenomenon may result from the substitution of La^{3+} for cation with a smaller radius such as bivalent cobalt. Accordingly, no shifts are found over $Co_{0.93}La_{0.07}$–pm sample. All the diffraction peaks of $Co_{0.93}La_{0.07}$ are

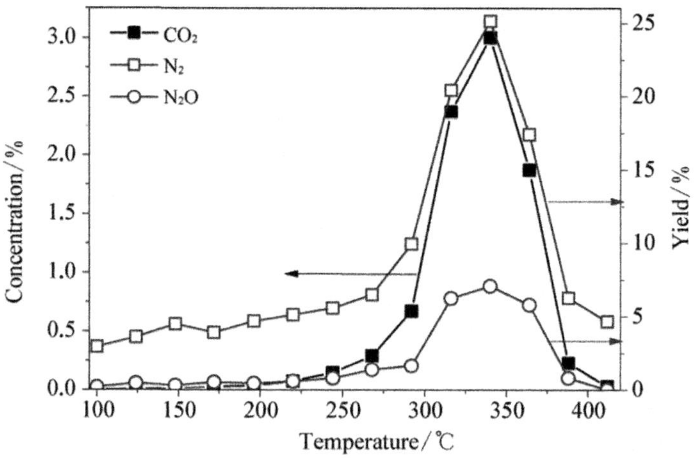

Fig. 2.29 Reaction profiles for CO_2 concentration, the yield of N_2 and yield of N_2O during simultaneous soot–NO_x removal process over $Co_{0.93}La_{0.07}$. Flow rate: 25 mL/min; gas: 2000 ppm NO + 5%O_2

Table 2.9 Summary of the catalytic performance for Co–La mixed oxides

Catalyst	T_{10} (°C)	T_{50} (°C)	T_{90} (°C)	Yield of N_2 (%) (340 °C)
Co_3O_4	315	378	425	13.0
$Co_{0.98}La_{0.02}$	293	329	364	22.5
$Co_{0.96}La_{0.04}$	289	330	370	23.2
$Co_{0.93}La_{0.07}$	286	325	369	25.2
$Co_{0.83}La_{0.17}$	290	336	372	24.2
$LaCoO_3$	290	339	374	23.2
$Co_{0.93}La_{0.07}$–pm	307	362	405	16
$LaCoO_3/SiO_2$ (physical mixture)	354	451	504	20.8

broadened, which revealed that crystal grain size of this sample is smaller than that of other catalysts.

H_2–TPR curves uncovering the redox ability of the catalysts are presented in Fig. 2.31. $LaCoO_3$ is characterized by two groups of reduction peaks. The first

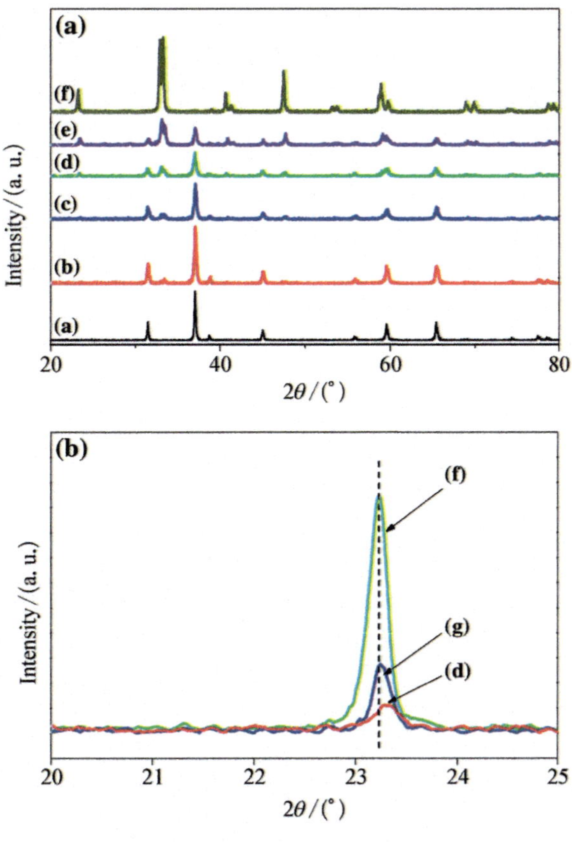

Fig. 2.30 XRD patterns (**A**) for $Co_{1-x}La_x$ catalysts and magnified patterns (**B**) for $LaCoO_3$ (012) [55]. (a) $x = 0$; (b) $x = 0.02$; (c) $x = 0.04$; (d) $x = 0.07$; (e) $x = 0.17$; (f) $x = 0.5$; (g) $Co_{0.93}La_{0.07}$–pm

group, occurring at 391 and 405 °C, is due to the one-electron reduction of Co^{3+}. $LaCoO_3$ is initially reduced to lanthanum hydroxide and metallic cobalt. The other group observed at 578 °C arises from reduction process of Co^{3+} to Co^0 [56]. The broadened and asymmetric shapes for Co_3O_4 imply that there exists more than one stage for the reduction of trivalent cobalt to metallic cobalt. The hydrogen uptake is assigned to two consecutive reduction steps, i.e., Co^{3+} to Co^{2+} and Co^{2+} to Co^0 [57]. The profile of $Co_{0.93}La_{0.07}$ and $Co_{0.93}La_{0.07}$–pm is similar to that of Co_3O_4 below 550 °C because the characteristics of $LaCoO_3$ were overlapped by a larger proportion of Co_3O_4. However, the difference between $Co_{0.93}La_{0.07}$ and $Co_{0.93}La_{0.07}$–pm is still distinguishable. A predominant peak at 370 °C indicates a stronger redox ability of $Co_{0.93}La_{0.07}$ than $Co_{0.93}La_{0.07}$–pm. Soot–TPR tests are applied to analyze the activity of surface oxygen and lattice oxygen (Fig. 2.32). Two species of active oxygen can be identified from DTG profiles. First being the surface adsorbed oxygen including O_2^- and O^- (500–720 °C), while the other being the lattice oxygen O^{2-} (>720 °C) [58]. A promotion effect is obtained over $Co_{0.93}La_{0.07}$ compared with its physical mixture. The surface oxygen, as well as the

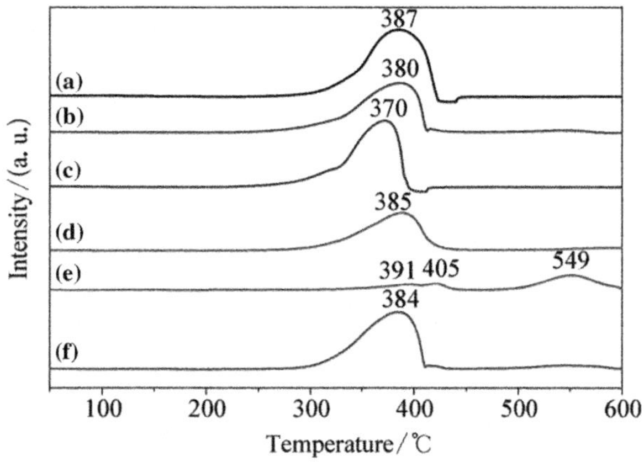

Fig. 2.31 H_2–TPR profiles for Co–La complex oxides. (a) Co_3O_4; (b) $Co_{0.96}La_{0.04}$; (c) $Co_{0.93}La_{0.07}$; (d) $Co_{0.83}La_{0.17}$; (e) $LaCoO_3$; (f) $Co_{0.93}La_{0.07}$–pm

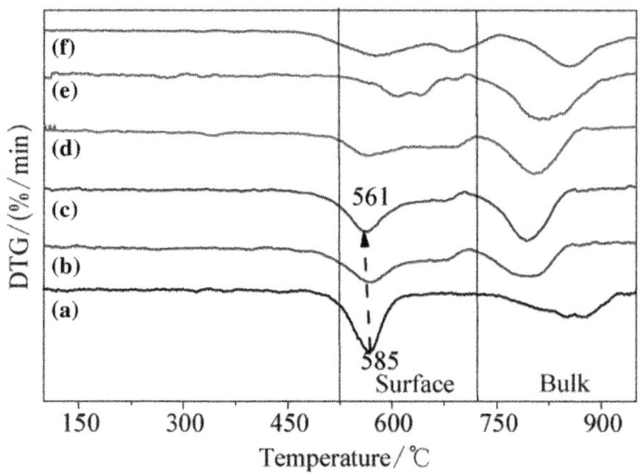

Fig. 2.32 Soot–TPR profiles for Co–La complex oxides. (a) Co_3O_4; (b) $Co_{0.96}La_{0.04}$; (c) $Co_{0.93}La_{0.07}$; (d) $Co_{0.83}La_{0.17}$; (e) $LaCoO_3$; (f) $Co_{0.93}Ce_{0.07}$–pm

lattice oxygen, is activated at a lower temperature. $Co_{0.93}La_{0.07}$ seemed to embrace the advantages of both materials.

XPS survey and its calculation results are demonstrated in Fig. 2.33 and Table 2.10. In Co $2p_{3/2}$ spectrum, shifts from 779.6 to 779.2 eV are also discovered in this study. It is known that the negative shifts of Co $2p_{3/2}$ peaks are due to the existence of Co^{4+} [59, 60]. This fraction of Co^{4+} is unlikely to originate from spinel Co_3O_4 since no low valence ions are introduced to the system to replace Co^{2+} or Co^{3+} in Co_3O_4. Co^{4+} possibly comes from the partial incorporation of Co^{2+} into A-site of $LaCoO_3$. The replacement of La^{3+} by Co^{2+} leads to the oxidation of B-site ions (Co^{3+} to Co^{4+}), and the formation of oxygen vacancies so that the electrical neutrality of $LaCoO_3$ can be maintained. Also, the proportion of oxygen adsorption slightly increases due to more oxygen vacancy in $Co_{0.93}La_{0.07}$ than that in $Co_{0.93}La_{0.07}$–pm.

Judging by the results above, it is postulated that a synergistic effect may occur between Co_3O_4 and $LaCoO_3$. The coexistence between the two oxides gives rise to the smaller size of crystal grain and the formation of Co^{4+}. The latter is one of the crucial factors for the promotion of the catalytic activity. In most cases, bivalent

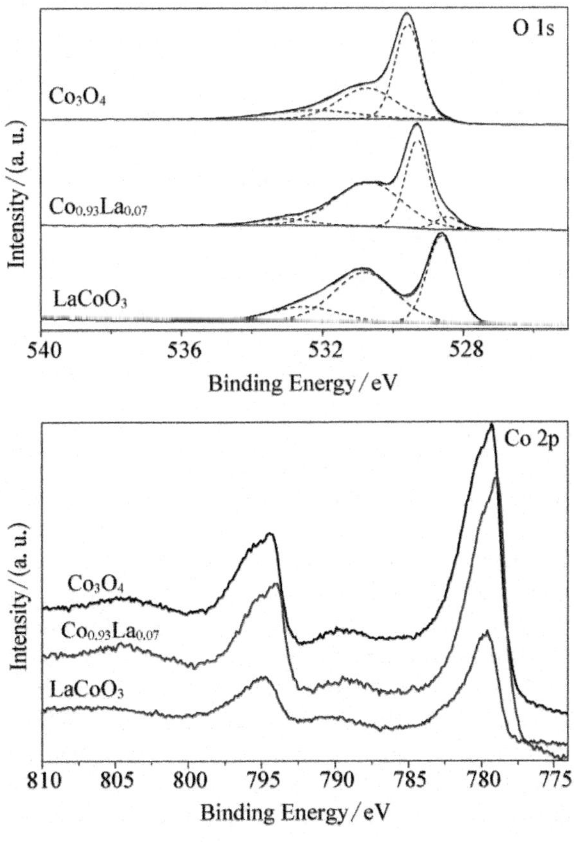

Fig. 2.33 XPS survey for O 1s and Co 2p for Co_3O_4, $Co_{0.93}La_{0.07}$, and $LaCoO_3$ [55]

Table 2.10 Summarized calculation for different kinds of oxygen species and binding energies [55]

Catalyst	O_I (%)	O_{II} (%)	Lattice oxygen[a] (%)	O_{III} (%)	O_{IV} (%)	Adsorbed oxygen[b] (%)	Co $2p_{3/2}$ (eV)
Co_3O_4	0	50.47	50.47	34.49	15.04	49.53	779.6
$Co_{0.93}La_{0.07}$–pm	9.66	34.44	44.1	50.0	5.9	55.9	779.5
$Co_{0.93}La_{0.07}$	5.37	37.27	42.64	51.27	6.09	57.36	779.0
$LaCoO_3$	38.92	0	38.92	46.33	14.75	61.08	779.3

cobalt hardly replaces metal ions or rare-earth ions at A-site of ABO_3 perovskite because it can easily occupy site B with higher oxidation state (Co^{3+}). In this case, Co^{2+} that replaces A-site in ABO_3 perovskite is provided by Co_3O_4. This crystal defect favors soot abatement and NO_x reduction.

2.4.3 Co–Ce Composite Oxides for Simultaneous Soot–NO_x Removal

Co–Ce composite oxides are also synthesized via CIT method. The cobalt acetate and cerium nitrate are used as precursors, and the final calcination temperature is 700 °C. The composite oxides are denoted as $Co_{1-x}Ce_x$ where x represents the molar ratio of Ce.

The reaction profiles for Co–Ce composite oxides are similar to that of Co–La ones, and the reaction conditions are also the same. The catalytic performances over Co–Ce composites are summed up in Table 2.11. In comparison with simple oxide (Co_3O_4 and CeO_2), the Co–Ce composite oxides show a promotional effect not only in the ignition temperature but also the N_2 conversion. It is discovered that T_{10} value exhibits an inverse correlation with cerium loading until $x = 0.07$; further introduction of cerium fails to lower the ignition temperature. In contrast, the optimized yield of N_2 is acquired over $Co_{0.83}Ce_{0.17}$ (the maximum cerium loading sample). Above all, the $Co_{0.93}Ce_{0.07}$ sample exhibited an excellent overall performance; T_{10} reaches 315 °C, and maximum yield of N_2 also achieves 29.6%. Moreover, it is observed that $Co_{0.93}Ce_{0.07}$ initiates soot combustion at a lower temperature than $Co_{0.93}Ce_{0.07}$–pm does, which indicates the possibility of an interaction between the two oxides. The following characterizations are referred to elucidate the brilliant performance of this material.

XRD patterns of Co–Ce mixed oxides catalysts are shown in Fig. 2.34a. Two crystal phases, namely Co_3O_4 with cubic spinel-type and CeO_2 with cubic fluorite structure, were detected over the compound oxides in the figure. Magnified XRD patterns for Co_3O_4 (311) are presented in Fig. 2.34b which illustrate the negative shifts occurred over Co_3O_4 (311). The maximum shift appears over $Co_{0.93}Ce_{0.07}$

Table 2.11 Summary of the catalytic performance for Co–Ce compound oxides in NO_x atmosphere

Catalyst	T_{10} (°C)	T_{50} (°C)	T_{90} (°C)	Yield of N_2 (%) (340 °C)
Co_3O_4	315	378	425	13.0
$Co_{0.98}Ce_{0.02}$	292	333	366	22.1
$Co_{0.96}Ce_{0.04}$	284	326	358	24.3
$Co_{0.93}Ce_{0.07}$	276	313	344	29.6
$Co_{0.83}Ce_{0.17}$	280	320	352	31.7
CeO_2	331	388	436	14.3
$Co_{0.93}Ce_{0.07}$–pm	303	350	430	20.0

catalysts among all the samples. Such shifts are observed for the whole Co_3O_4 pattern of $Co_{1-x}Ce_x$ samples. This probably results from the incorporation of cerium ions with larger radius into the Co_3O_4 lattice or the formation of a Co–Ce–O solid solution. This solid solution is believed to locate at the boundary of both oxides [61]. Considering the fact that no additional peak is distinguished apart from CeO_2 and Co_3O_4 peaks, the content of the solid solution is absolutely traced even if it

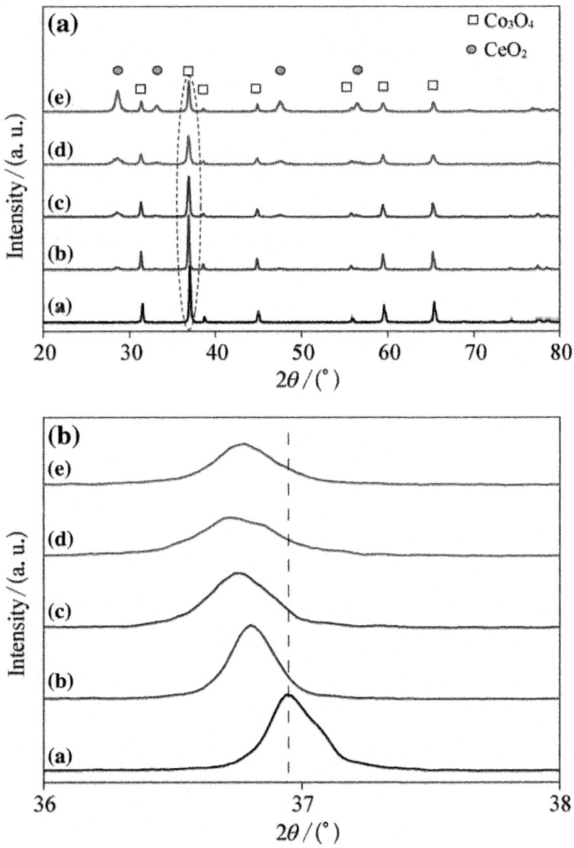

Fig. 2.34 XRD patterns of $Co_{1-x}Ce_x$ catalysts [62]. (**a**) whole pattern; (**b**) magnified pattern of Co_3O_4 (311); (a) $x = 0$; (b) $x = 0.02$; (c) $x = 0.04$; (d) $x = 0.07$; (e) $x = 0.17$

Table 2.12 Peak position and grain sizes of Co–Ce complex oxides acquired from XRD patterns

Catalysts	$2\theta°$ (Co_3O_4 (311))	$D_{Co_3O_4}$ (nm)	Co_3O_4 lattice parameter (0.1 nm)	D_{CeO_2} (nm)
Co_3O_4	36.880	36.8	8.06280	n/a
$Co_{0.98}Ce_{0.02}$	36.800	36.1	8.06787	16.1
$Co_{0.96}Ce_{0.04}$	36.740	24.5	8.07173	14.4
$Co_{0.93}Ce_{0.07}$	36.700	20.1	8.07780	10.0
$Co_{0.83}Ce_{0.17}$	36.78	22.2	8.06727	15.1
$Co_{0.93}Ce_{0.07}$–pm	36.88	36.7	8.06280	16.5
CeO_2	n/a	n/a	n/a	16.6

exists. Thus, it is likely that peak shifts of Co_3O_4 can largely be attributed to the incorporation of cerium into Co_3O_4. It is apparent that CeO_2 is responsible for the lattice expansion of Co_3O_4, and the radical expansion of lattice parameter is acquired over the $Co_{0.93}Ce_{0.07}$ catalyst corresponding to the best overall performance (Table 2.12).

The Raman spectroscopic technique was adopted to explore the structure of catalysts, and the spectra are given in Fig. 2.35. The bands at 692 cm^{-1} (A_{1g} mode), 620 cm^{-1} (F_{2g}^3 mode), 520 cm^{-1} (F_{2g}^2 mode), and 482 cm^{-1} (E_g mode) are assigned to spinel-type Co_3O_4 [63]. Pronounced redshifts of all Co_3O_4 peaks are observed along with the variation of cerium contents. Given the fact that the phonon confinement effect was negligible in Co_3O_4 [64], the redshifts of Raman spectrum are concluded to be the results of the lattice distortion. The maximum offset value is viewed over $Co_{0.93}Ce_{0.07}$, which implies this sample experienced extreme lattice distortion. Information acquired from Raman spectrum coincides with calculation results of the lattice parameter (Table 2.12). Nevertheless, characteristic peaks of the fluorite-type structure CeO_2 (464 cm^{-1}) are hardly observed in most samples

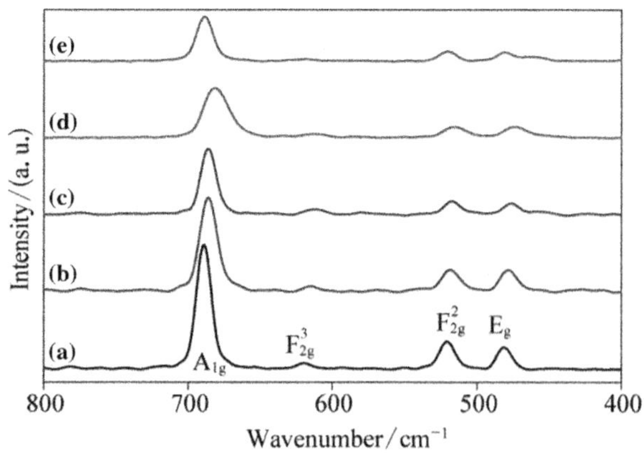

Fig. 2.35 Raman spectrum of $Co_{1-x}Ce_x$ [62]. (a) $x = 0$; (b) $x = 0.02$; (c) $x = 0.04$; (d) $x = 0.07$; (e) $x = 0.17$

except $Co_{0.83}Ce_{0.17}$ [58]. This is probably ascribed to the variation of CeO_2 content in the catalyst.

H_2–TPR (Fig. 2.36) and soot–TPR (Fig. 2.37) reveal that the Co–Ce composite oxides favor the redox ability and the activity of lattice oxygen for the catalyst, respectively. Among all the samples, $Co_{0.93}Ce_{0.07}$, corresponding to best overall catalytic activity, is able to consume hydrogen in H_2–TPR and to activate lattice in soot–TPR at a rather low temperature. This suggests that the reaction is associated with the redox properties of catalysts.

NO_x–TPD was performed to study the surface adsorption of NO species on the catalysts in NO_x (Fig. 2.38 and Table 2.13). Previous research pointed out that the thermal stability of NO_x adsorption species follows the sequence: nitrite ions > nitrate ions > surface nitrosyls (50 °C). The profiles consist of four peaks, corresponding to different temperatures: low (150 °C), middle (230 °C), and high (330 and 410 °C). The low-temperature peak and the middle-temperature one

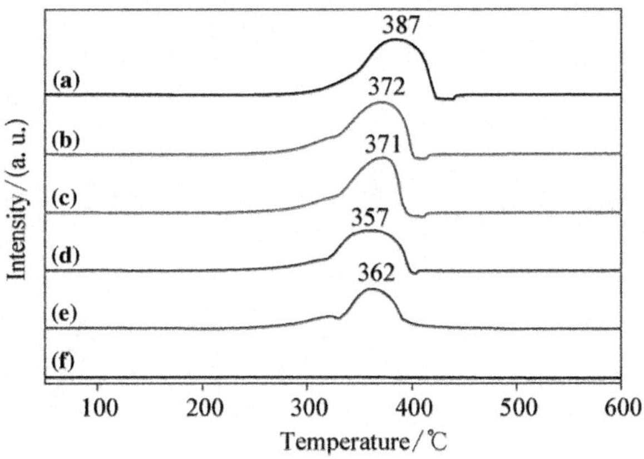

Fig. 2.36 H_2–TPR profiles for (a) Co_3O_4, (b) $Co_{0.98}Ce_{0.02}$, (c) $Co_{0.96}Ce_{0.04}$, (d) $Co_{0.93}Ce_{0.07}$ (e) $Co_{0.83}Ce_{0.17}$ and (f) CeO_2 [62]

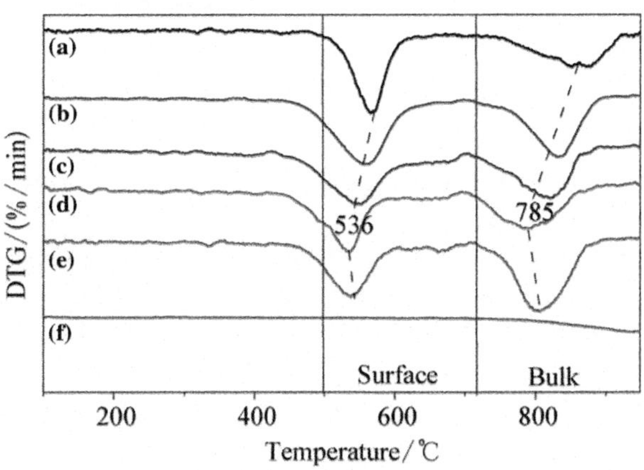

Fig. 2.37 Soot–TPR profiles for Co–Ce complex oxides [62]. (a) Co_3O_4; (b) $Co_{0.98}Ce_{0.02}$; (c) $Co_{0.96}Ce_{0.04}$; (d) $Co_{0.93}Ce_{0.07}$; (e) $Co_{0.83}Ce_{0.17}$; (f) CeO_2

corresponded to decomposition or desorption of nitrosyl and nitrite, respectively. The double peaks located at higher temperatures are ascribed to nitrate bonded to cobalt and cerium in sequence [65]. In general, the NO_x adsorption capability within the preset temperature range for Co_3O_4 is poor. In contrast, a massive amount of NO_x can adhere to CeO_2; a majority of adsorbate releases at the middle temperature (200–300 °C). Co–Ce composite oxides, represented by $Co_{0.93}Ce_{0.07}$, further improve the NO_x adsorption and desorption ability. It is also highlighted that there is 16 μmol/g gap between $Co_{0.93}Ce_{0.07}$ and its physical mixture ($Co_{0.93}Ce_{0.07}$–pm), which indicates that the interaction between Co_3O_4 and CeO_2 plays a crucial role in the NO_x-assisted reaction. The NO_x adsorption for Co–Ce composites is contributed by three portions: Co_3O_4, CeO_2 and the interaction between the previous two oxides. Among all the Co–Ce oxides, the maximum NO_x desorption value is acquired over $Co_{0.83}Ce_{0.17}$ which is dominant in the yield of N_2. This phenomenon demonstrates that the NO_x reduction process is likely to be associated with NO_x adsorption and desorption sites.

The XPS survey and its calculation results are illustrated in Fig. 2.39 and Table 2.14. All samples are characterized by three peaks in O 1s spectra. The peak (O_I) with lower binding energy (~529.6 eV) was assigned to the lattice oxygen [66]. The shoulder peak (O_{II}) exhibiting higher binding energy (~531.7 eV)

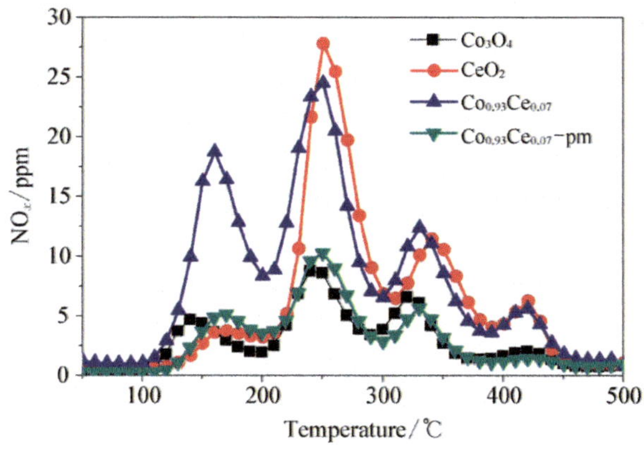

Fig. 2.38 NO–TPD profiles of NO_x ($NO + NO_2$) for Co_3O_4, CeO_2, $Co_{0.93}Ce_{0.07}$, and $Co_{0.93}Ce_{0.07}$–pm

Table 2.13 Summarized results of Co–Ce complex oxides for desorption amount of NO, NO_2, and NO_x

Catalyst	NO (μmol g)	NO_2 (μmol g)	NO_x^a (μmol g)
Co_3O_4	8.74	1.68	10.42
$Co_{0.96}Ce_{0.04}$	24.10	4.70	28.80
$Co_{0.93}Ce_{0.07}$	24.31	4.13	28.44
$Co_{0.83}Ce_{0.17}$	20.98	8.75	29.73
CeO_2	19.86	1.86	21.72
$Co_{0.93}Ce_{0.07}$–pm	9.4	3.04	12.44

corresponds to the adsorbed oxygen species (O^- or –OH) [67–69]. In other words, O_{II} was the sign of the chemisorbed oxygen. O_{III} (~532.7 eV) is ascribed to weakly bound surface oxygen [70]. The proportion of each oxygen species is listed in Table 2.14. It is found that the ratio of the adsorbed oxygen to the surface lattice oxygen is just about the same level for two samples. However, the total amount of oxygen in the surface exhibited distinct differences. The molar ratio of oxygen to total metal atom achieved 1.70 for Co_3O_4, 2.53 for $Co_{0.93}Ce_{0.07}$ and 3.57 for $Co_{0.83}Ce_{0.17}$, respectively. This ratio of all materials exceeds the stoichiometric value of Co_3O_4. Further, the value of Co–Ce mixed oxides samples even surpasses the stoichiometric value of CeO_2, which indicates the exposure of tremendous amount of oxygen in the surface. Although the percentage of adsorbed oxygen in the surface of Co_3O_4 and $Co_{0.93}Ce_{0.07}$ is similar, a larger absolute amount of oxygen is acquired over $Co_{0.93}Ce_{0.07}$. On the basis of this fact, $Co_{0.93}Ce_{0.07}$ shows the best ability to adsorb oxygen. It is inferred that the oxygen adsorption capability may contribute to its excellent activity. Apart from the information in O 1s spectrum, it is noticed that the cerium-containing samples experienced negative shifts with the Co $2p_{3/2}$ peak from 779.3 to 778.9 eV. This phenomenon can be attributed to the existence of higher valence state of cobalt ions such as Co^{4+} and/or Co^{3+}. Furthermore, a weak peak (~904 eV) in Ce 3d spectrum implies the appearance of Ce^{3+} [71, 72].

CeO_2 confirmed in XRD patterns is likely to play several roles in the reaction. Firstly, CeO_2 itself is a highly active material and can provide an extra source of oxygen including adsorbed oxygen and lattice oxygen in the surface and the bulk. Adsorbed oxygen is likely to be dissociated into superoxide followed by further oxidizing nitrite or nitrate species and attacking soot. Meanwhile, active lattice oxygen will take part in the reaction at the three-phase boundary [51]. Secondly, CeO_2 also contributes to oxygen molar ratio to the metal ions of the catalyst due to its higher stoichiometric value than that of Co_3O_4.

In fact, this is one factor but not the crucial one. The inferior performance of $Co_{0.93}Ce_{0.07}$–pm to that of $Co_{0.93}Ce_{0.07}$ uncovers there may be other determinants. The interaction between the two oxides can be a beneficial factor for the distinctions of catalytic behaviors to a great extent. The interaction can be divided into two parts, or rather, the lattice distortion of Co_3O_4 and the formation of Co–Ce solid solution. The XPS survey provides some clues for the existence of Co–Ce solid solution. The introduction of cerium cations into Co_3O_4 leads to the formation of more cobalt cations with higher valence state (Co^{3+} and Co^{4+}) and a small fraction of trivalent cerium. It is unlikely to be the effect of cations replacement. The replacement of cobalt cations by cerium cations with lower valence state may give rise to the formation of cobalt ions with higher valence, and more oxygen vacancies so that the electrical neutrality of Co_3O_4 can be maintained. Unfortunately, cerium ions in this study existed in the form of Ce^{3+} and Ce^{4+}, which shows a higher valence state. The existence of higher valence cobalt, therefore, may be the results of solid solution formation. Although the Co–Ce solid solution may modify the properties of catalysts, its trace amount is unlikely to be the major part of the interaction. Further, there is little possibility for the soot to contact Co–Ce solid

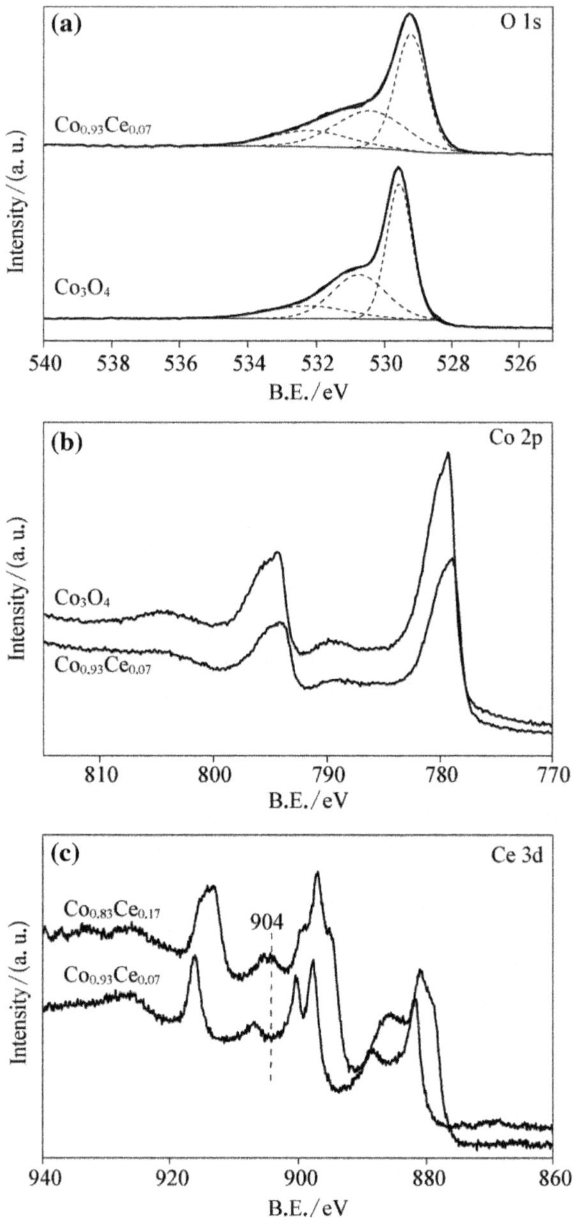

Fig. 2.39 XPS spectra for **a** O 1s, **b** Co 2p and **c** Ce 3d [62]

solution at the boundary of Co_3O_4 and CeO_2; hence, the solid solution almost has no effects on soot combustion.

CeO_2 plays a role modifying the Co_3O_4 structure. The analysis of XRD and Raman spectrum reveals the incorporation of cerium ions into the Co_3O_4 lattice

Table 2.14 Statistics of XPS survey [62] (Fig. 2.39)

Catalysts	O_I (%)	O_{II} (%)	O_{III} (%)	Adsorbed oxygen (%)[a]	$O^*/(Co^* + Ce^*)$[b]	Co $2p_{3/2}$ (eV)
Co_3O_4	50.47	34.49	15.04	49.53	1.70	779.6
$Co_{0.93}Ce_{0.07}$	49.00	34.62	16.38	51.00	2.53	779.1
$Co_{0.83}Ce_{0.17}$	76.00	9.68	14.32	24.00	3.57	778.8

[a]Adsorption oxygen percentage equals to $O_{II} + O_{III}$
[b]$O^*/(Co^* + Ce^*)$ represents the overall ratio of oxygen atoms to total metal atoms

which probably caused by the profound effects of cerium cations and cobalt cations. A similar phenomenon was also reported that the cerium ions were incorporated into the lattice of La_2O_3 and other spinel-type materials [73, 74]. The level of incorporation can be expressed by the lattice parameter of Co_3O_4 which exhibits a positive relationship with catalytic activities (Table 2.11 and 2.12). Actually, the catalytic performances, brought about by the lattice distortion, are consequently controlled by the modification of the surface and the lattice. The NO_x desorption performance is promoted over mixed oxides. The NO_x adsorption ability is also slightly intensified by additional oxygen vacancies created by incorporation. The interaction contributes to improving both the NO_x adsorption and desorption process. It is postulated that NO_x adsorption and desorption are the critical processes for the acceleration of soot oxidation. In addition to the surface change, CeO_2 also affects the lattice oxygen of the mixed oxides. This lattice distortion brings about the lower activation temperature of lattice oxygen. The interaction between two oxides, resulting from structure change, leads to the improvement of the redox ability, and this is well illustrated in H_2–TPR tests. It is surprising that the Ce^{3+} remains stable after high-temperature calcination in rich oxygen, which also implies the strong interaction between Co_3O_4 and CeO_2.

2.5 Ordered Mesoporous Oxide Material for Simultaneous Soot–NO_x Removal

Ordered mesoporous catalyst allows the properties of materials to be manipulated in a controlled manner, such as tunable composition, controllable structure, and morphology [75, 76]. Therefore, it opens the door to new catalytic processes and routes to increase the catalytic activity and selectivity for specific reaction [77–80]. However, by far the most mentioned porous architectures applied in soot combustion were macroporous materials [81, 82], primarily due to considering the limited mesopore size compared with soot particles, yet few examples were reported to use ordered mesoporous materials for soot–NO_x simultaneously removal reaction. The mesopores can ensure optimal accessibility and transport of gas-phase reactants and products, such as NO_x and C_xH_yO [78, 83–85]. Such unique structure properties may also strengthen the contact of soot and catalyst and transport of gas-phase reactant and possibly improve the performance in NO_x–soot removal reaction. It is therefore highly desirable to investigate the dependence of the properties of mesoporous materials on influencing the NO_x–soot removal catalytic activity.

2.5.1 Catalytic Performance of Mesoporous Co_3O_4

Co_3O_4 was selected as a model catalyst since spinel-type catalysts were already proved to be good candidates for the NO_x–soot catalysis removal reaction [5, 86]. Mesoporous Co_3O_4 was synthesized through nanocasting method using SBA-15 as hard template, methanol as a dispersing agent and NaOH or HF to remove the template. The mesoporous Co_3O_4 using NaOH to remove template was denoted as m–Co_3O_4–Na, and the mesoporous Co_3O_4 using HF to remove template was denoted as m–Co_3O_4–F. The mesoporous Co_3O_4 using KOH to remove template was denoted as m–Co_3O_4–K. Addition mesoporous Co_3O_4 using NaOH to remove template replicated from KIT-6 was denoted as k–Co_3O_4–Na. As a comparison, b–Co_3O_4–S was bulk Co_3O_4 synthesized through sol–gel method.

A typical profile for simultaneous soot–NO_x removal over b–Co_3O_4–S was presented in Fig. 2.40. Three products, i.e., nitrogen (N_2), nitrous oxide (N_2O), and carbon dioxide (CO_2), were detected during the reaction. In the reaction temperature ranged from 100 to 400 °C, all products exhibited "volcano-shape" behavior in soot–NO_x removal reaction, namely the conversion efficiency increased until the maximum (317 °C) was reached and then decreased due to the run out of the charged soot. CO_2 was formed through the oxidation of soot. Although soot may be oxidized by NO_x or O_2, the simultaneous detection of CO_2, N_2, and N_2O at the same temperature range together with the similar tendency of the evolution profile indicated that soot was consumed by NO_x rather than O_2 [5, 13].

The catalytic performance was evaluated through the following parameters. The ignite temperature of soot (T_{10}) was defined as the temperature of 10% formation of carbon dioxide for no carbon monoxide was detected. The 50% soot combustion temperature (T_{50}) was defined as the temperature of 50% formation of carbon dioxide. The peak temperature (T_p) was defined as the peak temperature in the CO_2-temperature profile. Total amounts of N_2, N_2O, and CO_2 formed throughout the reaction were obtained by integrating the respective conversion versus temperature (time) curves. $S_{N_2/C}$ was defined to evaluate the proportion of soot used for the reduction of NO_x into nitrogen, which is equal to the molar ratio of the formed nitrogen to the charged soot or carbon dioxide formed. Also, S_{N_2O} and S_{N_2} defined as the selectivity to nitrous oxide were calculated by the following equations: $S_{N_2O} = V[N_2O]/(V[N_2O] + V[N_2])$, $S_{N_2} = V[N_2]/(V[N_2O] + V[N_2])$.

A comparison study was carried out to examine the activity of the following three Co_3O_4 samples for the soot and NO_x simultaneous removal reaction

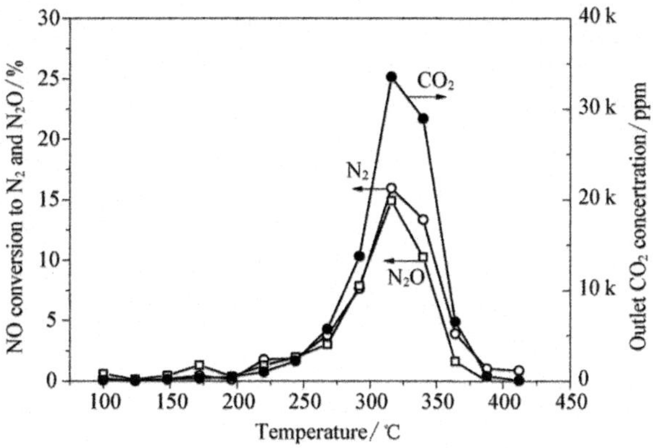

Fig. 2.40 Temperature dependence of NO_x conversion into N_2, N_2O and CO_2 formed throughout TPO run in the simultaneous removal of NO_x and soot over b–Co_3O_4–S [87]

(Fig. 2.41). The evaluate parameters are illustrated in Table 2.15. It can be seen that b–Co_3O_4–S showed the lowest activity, giving highest T_{10}, T_{50} (276, 321 °C) and lowest $S_{N_2/C}$ (0.6%). The conversion ratio of NO to N_2 at T_p (317 °C) is only about 16% (Fig. 2.41) while m–Co_3O_4–Na exhibited the best performance among the three samples. It could achieve the lowest T_{10} and T_{50} (250, 286 °C), which means the ignite temperature of m–Co_3O_4–Na decreased about 26 °C compared with b–Co_3O_4–S. The conversion ratio of NO to N_2 at T_p (292 °C, Fig. 2.41) reached 33% for m–Co_3O_4–Na in accordance with the highest value of $S_{N_2/C}$ (1.2%) which was about twice that of b–Co_3O_4–S (16 and 0.6%, respectively). High value of $S_{N_2/C}$ indicated more nitrogen yielded over m–Co_3O_4–Na, which suggested NO_x was reduced readily into N_2 over m–Co_3O_4–Na. As the selectivity to N_2O exhibited no obvious distinction over all three samples (Table 2.15), the above results indicated that m–Co_3O_4–Na improved the formation of nitrogen and nitrous oxide rather than more N_2 with less N_2O. Another mesoporous Co_3O_4 sample, m–Co_3O_4–F, only showed slight higher selectivity to N_2 compared with b–Co_3O_4–S. The conversion ratio of NO to N_2 at T_p and $S_{N_2/C}$ was 21 and 0.7%, respectively, which was about 30 and 17% higher than that of b–Co_3O_4–S. But it was still 51 and 71% lower than that of m–Co_3O_4–Na. The ignite temperatures of m–Co_3O_4–F and b–Co_3O_4–S are very close, higher than that of m–Co_3O_4–Na. The striking difference in the catalysis performance among the three Co_3O_4 catalysts may originate from the following structure properties: (i) texture properties; (ii) morphology properties; (iii) surface properties. The impacts of synthesis condition on mesoporous and bulk Co_3O_4 catalysts were undertaken to elucidate which factor was responsible for the enhanced performance.

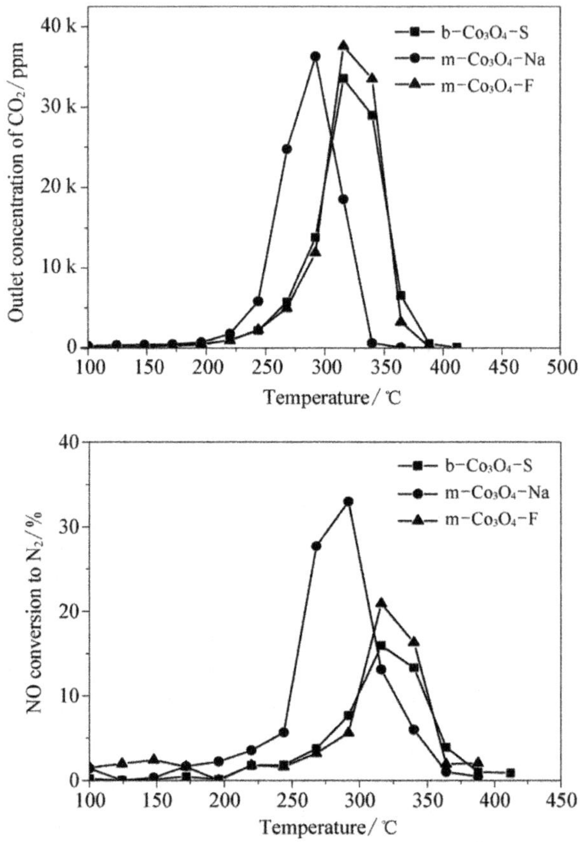

Fig. 2.41 Catalytic performance of the Co_3O_4-bulk, m–Co_3O_4–Na, and m–Co_3O_4–F [87]

Table 2.15 Catalytic performance of the samples [87]

Catalysts	T_{10} (°C)	T_{50} (°C)	$V[N_2]$ (10^{-5} mol)	Selectivity (%)		
				$S_{N_2/C}$	S_{N_2}	S_{N_2O}
b–Co_3O_4–S	276	321	1.22	0.6	56.6	43.4
m–Co_3O_4–Na	250	286	2.31	1.2	56.6	43.5
m–Co_3O_4–F	278	320	1.43	0.7	55.0	45.0

2.5.2 *Insight into Mechanism for Co_3O_4 Mesoporous Material*

The structure and texture properties of the catalysts were characterized by XRD and N_2-physisorption. Low-angle XRD pattern of the template SBA-15 and the as-prepared cobalt oxide samples were shown in Fig. 2.42a. Low-angle XRD patterns indicated Co_3O_4 prepared by nanocasting method possess the expected mesoporous structure. For the hard template SBA-15, three well-resolved diffraction peaks were detected at 0.9°, 1.6°, and 1.8°, corresponding to (1 0 0), (1 1 0), and (2 2 0) peak, suggesting 2D hexagonal mesoporous structure with a p6 mm space group for hard template SBA-15 [88]. For m–Co_3O_4–F and m–Co_3O_4–Na, peaks at 0.9° and 1.6° presented the (100) and (110) reflection were both found in the low-angle XRD pattern in Fig. 2.42a, suggesting the existence of the ordered mesoporous structure [89]. Figure 2.42b shows the wide-angle XRD pattern of the samples. By comparing with the standard data of Co_3O_4 (JCPDS PDF# 42-1467), it showed that the three samples were all crystalline Co_3O_4 with spinel-type structure as the samples displayed same diffraction peaks at 19°, 31.3°, 36.9°, 38.2°, 44.5°, 55.6°, 59.4°, and 65.3°, corresponding to the (1 1 1), (2 2 0), (3 1 1), (2 2 2), (4 0 0), (4 2 2), (5 1 1), and (4 4 0) planes [90]. Compared with b–Co_3O_4–S, the two mesoporous Co_3O_4 samples both showed weaker and broader diffraction peaks. It indicated that the two mesoporous Co_3O_4 both had smaller grain size, mainly caused by the restriction of the tunnels during the grain growth [91].

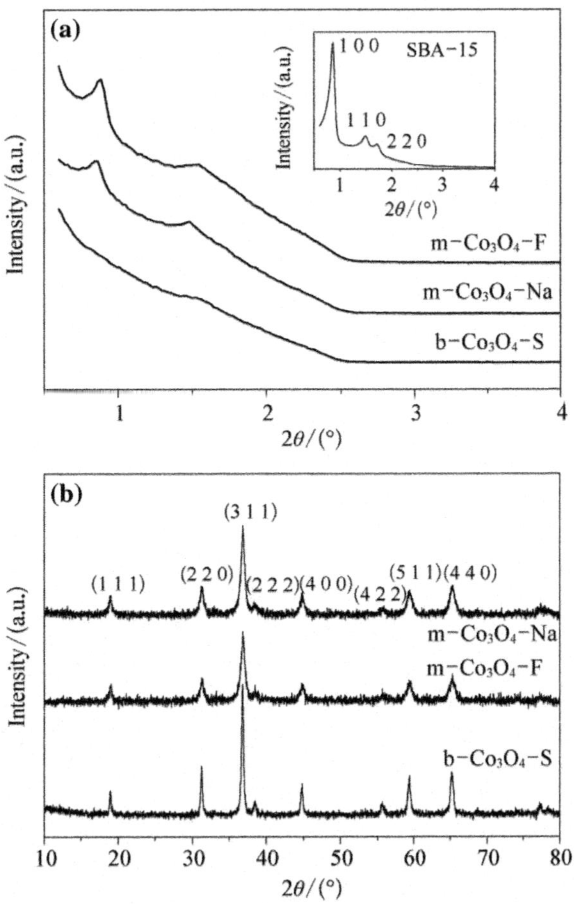

Fig. 2.42 Low-angle (**a**) of SBA-15 and mesoporous Co_3O_4 and wide-angle (**b**) of Co_3O_4 samples [87]

The textural properties of these catalysts were studied by N_2 physical adsorption–desorption. As shown in Fig. 2.43, the N_2 physisorption isotherms of these two mesoporous Co_3O_4 samples showed typical type-IV curves with hysteresis loops, suggesting the existence of mesopores [89]. In regard to the shape of the hysteresis ring, the mesoporous Co_3O_4 samples were not like that of the hard template SBA-15, which may be associated with interparticle voids [92]. The capillary condensation was observed in the relative pressure ranging from 0.4 to 0.8 for both mesoporous samples. The BJH pore size distributions of mesoporous materials showed that the m–Co_3O_4–Na and m–Co_3O_4–F had the mesoporous diameters of maximum distribution, 3.6, and 3.4 nm, respectively. This further indicated that m–Co_3O_4 independent of the template removal procedure was both perfect replica structures of their templates. The textural properties for m–Co_3O_4–Na and m–Co_3O_4–F were summarized in Table 2.16. They showed negligible difference in specific surface area, pore size, and pore volume. The surface area and pore volume of m–Co_3O_4–Na and m–Co_3O_4–F were about 90 m^2/g and 0.12–0.13 cm^3/g, which were higher than that of b–Co_3O_4–S (22 m^2/g, 0.08 cm^3/g).

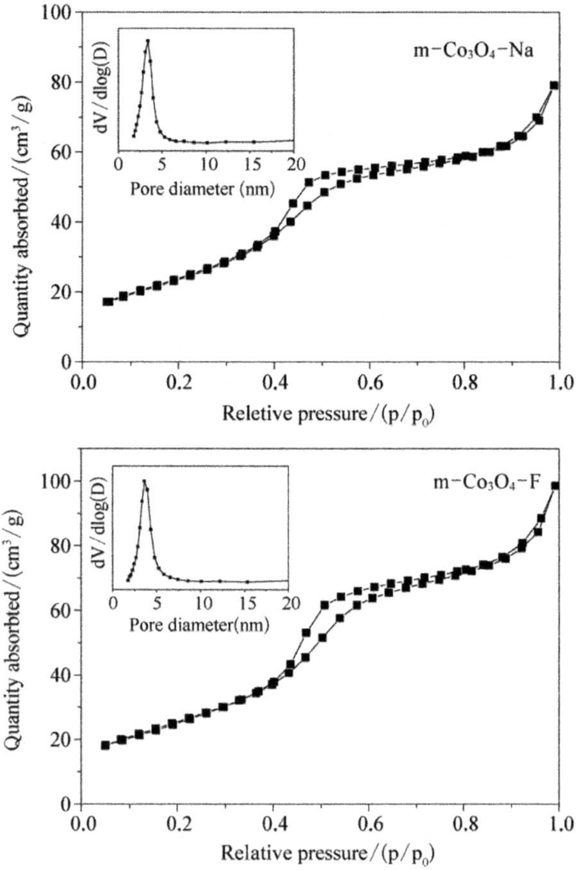

Fig. 2.43 N_2 physisorption isotherms and pore size distribution curves of the samples [87]

Table 2.16 Textural properties of b–Co_3O_4–S, m–Co_3O_4–F, and m–Co_3O_4–Na [87]

Sample	BET surface area (m^2 g)	Pore volume (cm^3 g)	Average pore diameter (nm)
b–Co_3O_4–S	22	0.08	–
m–Co_3O_4–Na	91	0.12	4.36
m–Co_3O_4–F	97	0.13	4.83

Figure 2.44 shows the TEM images of the Co_3O_4 catalysts. As shown in Fig. 2.44a–c, b–Co_3O_4–S showed non-porous structures with the particles size around 30 nm which was corresponding with the low-angle XRD results (Fig. 2.42a). The lattice fringes could be clearly observed in the HRTEM image

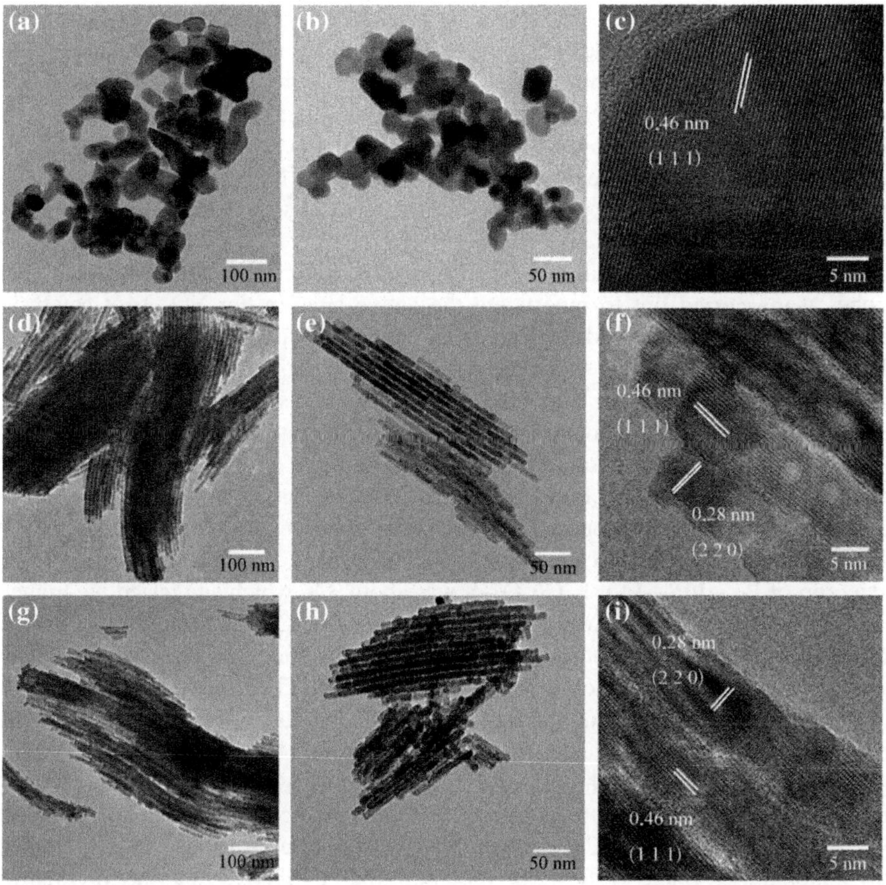

Fig. 2.44 TEM images of b–Co_3O_4–S (**a**, **b**, **c**), m–Co_3O_4–Na (**d**, **e**, **f**), m–Co_3O_4–F (**g**, **h**, **i**) Reproduced from Ref. [87] with permission from The Royal Society of Chemistry

(Fig. 2.44c) suggesting highly crystalline nature of the nanoparticles. The HRTEM image showed the b–Co_3O_4–S had the (1 1 1) crystal plane with a lattice spacing of 0.46 nm. Although the morphology of b–Co_3O_4–S was entirely different from that of m–Co_3O_4, TEM and HRTEM results did not support morphology difference for m–Co_3O_4–F and m–Co_3O_4–Na. The two mesoporous samples both possessed highly ordered 2D porous structure with polycrystalline walls according to Fig. 2.44d–i, suggesting m–Co_3O_4 retained the ordered structure of the SBA-15 template, which was consistent with the results of low-angle XRD and N_2 physisorption. The walls of mesoporous Co_3O_4 were both about 9 nm that was much smaller the Co_3O_4-bulk.

According to the HRTEM images, the mesoporous samples both had the (1 1 1) crystal plane with a lattice spacing of 0.46 nm and the (2 2 0) crystal plane with a lattice spacing of 0.28 nm [93]. The (1 1 1) crystal plane of b–Co_3O_4–S was suggested not be an active surface because the (0 0 1) and (1 1 1) faces mainly contained Co^{2+} [94–96], whereas the (2 2 0) crystal plane in parallel with the (1 1 0) crystal plane was an active surface as more active Co^{3+} species on the (1 1 0) faces, thus provided favorable sites for oxygen adsorption in catalytic reactions. This seems to be able to explain the inferior catalysis activity of b–Co_3O_4–S showed compared with m–Co_3O_4–Na in the NO_x–soot simultaneously removal reaction. But this still could not explain why only m–Co_3O_4–Na exhibited superior catalysis activity while m–Co_3O_4–F only showed slightly higher NO to N_2 conversion ratio than that of b–Co_3O_4–S as well as no improvement in soot combustion compared with b–Co_3O_4–S.

Temperature-programmed experiments using different chemical species as probes were carried out to investigate the active sites in the NO_x–soot catalysis reaction further. Thermal redox behavior of different Co_3O_4 was investigated using H_2–TPR. As shown in Fig. 2.45a, all three samples showed two reduction range, the low-temperature section peaks at 294 °C could be ascribed to the reduction of Co^{3+} to Co^{2+} as an intermediate while the high-temperature section was related to the reduction of Co^{2+} to Co^0 [97, 98]. The peak areas of P_{H2}-I and P_{H2}-II were summarized and showed in Fig. 2.45b. The peak area ratio of P_{H_2}-II/P_{H_2}-I for two mesoporous Co_3O_4 was around 3.5, which was approximately quantitatively consistent with the theoretical calculation value (P_{H_2}-II/P_{H_2}-I = 3) [97]. But the peak area ratio of P_{H_2}-II/P_{H_2}-I for b–Co_3O_4–S was about 3.8, which suggested the higher amount of Co^{2+} in the b–Co_3O_4–S compared with m–Co_3O_4. Thus, the H_2–TPR results indicated that mesoporous structure of Co_3O_4 could improve the reduction ability of Co^{3+} to Co^{2+} which was consistent with HRTEM results.

Temperature-programmed oxidation (TPO) of soot was further carried out using thermogravimetric analysis to investigate the oxidation behavior of soot under an oxygen atmosphere. As shown in Fig. 2.46, m–Co_3O_4–Na showed expected improved activity (decreasing T_{ig} about 25 °C) for soot combustion compared with b–Co_3O_4–S. However, m–Co_3O_4–Na only showed slightly higher activity in the low-temperature range (100–350 °C), but the derivative thermogravimetric (DTG) peak of m–Co_3O_4–F overlapped with m–CO_3O_4–Na at 375 °C. This was

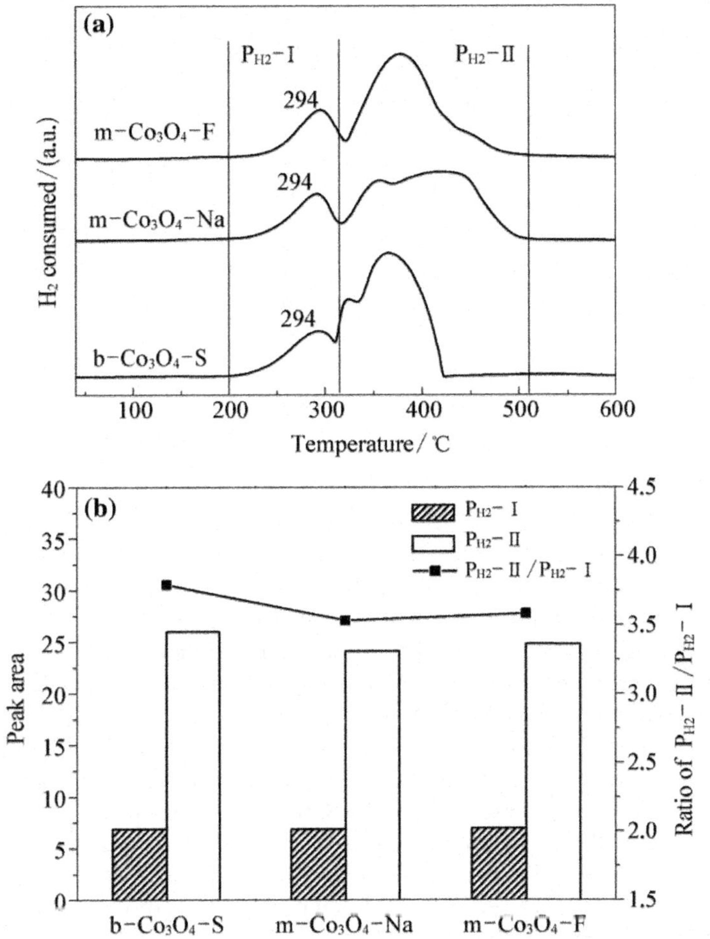

Fig. 2.45 TPR profiles of the samples (**a**) and peak area of Co_3O_4 catalysts (**b**) [87]

consistency with H_2–TPR and HRTEM results as which already suggested similar oxidation ability of the two mesoporous Co_3O_4. However, it is worthwhile mentioning that soot TPO under oxygen showed discrepant results compared with the reported promotion effect of m–Co_3O_4–Na on ignition of soot, while m–Co_3O_4–F displayed similar ignition properties as b–Co3O4–S under NO_x–O_2 atmosphere, suggesting the possible existence of balance between reaction conditions, state of doping element, contact conditions together with surface oxidation reducibility. Thus, soot oxidation pathway should not be the rate-determining step of the selective reduction of NO by soot over Co_3O_4.

Since the active oxygen species play a less prominent role in soot and NO_x simultaneously removal reaction under NO_x–O_2, NO_x–TPD was performed to investigate the NO_x absorption–desorption behavior of the different Co_3O_4 samples.

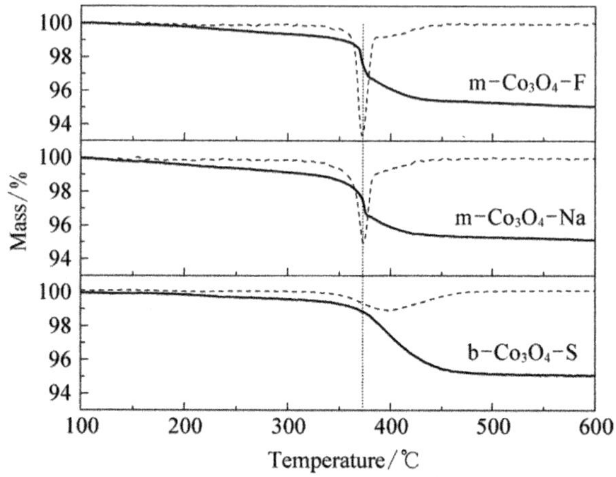

Fig. 2.46 TPO analysis profiles of Co_3O_4 samples [87]

TPD profiles (concentration versus temperature) of Co_3O_4 after saturation with 2000 ppm NO and 5% O_2 at 40 °C were shown in Fig. 2.47. In the case of the b–Co_3O_4–S sample, two NO desorption peaks were observed. The first one was in the range 100–200 °C (of lower intensity) with a maximum around 135 °C and the second appeared between 200 and 400 °C with a maximum near 270 °C accompanied with a shoulder near 340 °C. NO_2 desorption profile showed similar spectrum, with two NO_2 desorption peaks, the low-temperature peak centered at 131 °C, while the second stage (200–400 °C) centered at 244 °C. Although NO_x desorption amount was significantly higher than b–Co_3O_4–S, the m–Co_3O_4–F sample apparently displayed a similar trend as b–Co_3O_4–S, with NO peaks observed near 131 and 306 °C together with NO_2 peaks centered at 150 and 256 °C, respectively. In the case of NO, the lower desorption peak could be ascribed to the removal of nitrites and/or NO–$NO_2^{\delta+}$ surface species, and the high-temperature desorption peak was likely associated with the decomposition of nitrites species [98, 100]. With respect to NO_2, the peaks in the 100–200 °C region could be assigned to the removal of weakly absorbed NO_2–N_2O_4 and/or $NO_2^{\delta+}$ species, while the high-temperature peak may arise from the decomposition of nitrates intermediates [101, 102].

The m–Co_3O_4–Na samples exhibited different features compared with m–Co_3O_4–F and b–Co_3O_4–S, in aspects on the one hand to the presence of NO desorption peaks at high temperature (up to 500 °C) and on the other hand to the significantly higher overall amounts of NO_2 released especially at the second temperature range(200–400 °C). The NO peak of m–Co_3O_4–Na at high temperature could be related to the decomposition of NO_2 or ionic Na nitrites, the latter showing higher thermal stability [103].

NO–TPO was further performed to evaluate the oxidation behavior of NO using different Co_3O_4. Figure 2.48 shows the NO_2 concentration during TPO run. For b–Co_3O_4–S and m–Co_3O_4–F, two stages of NO_2 formation occurred with the increase of temperature. The first peak in the range 100–200 °C centered around 120 °C, which was consistent with the first NO_2 desorption peak (100–200 °C)

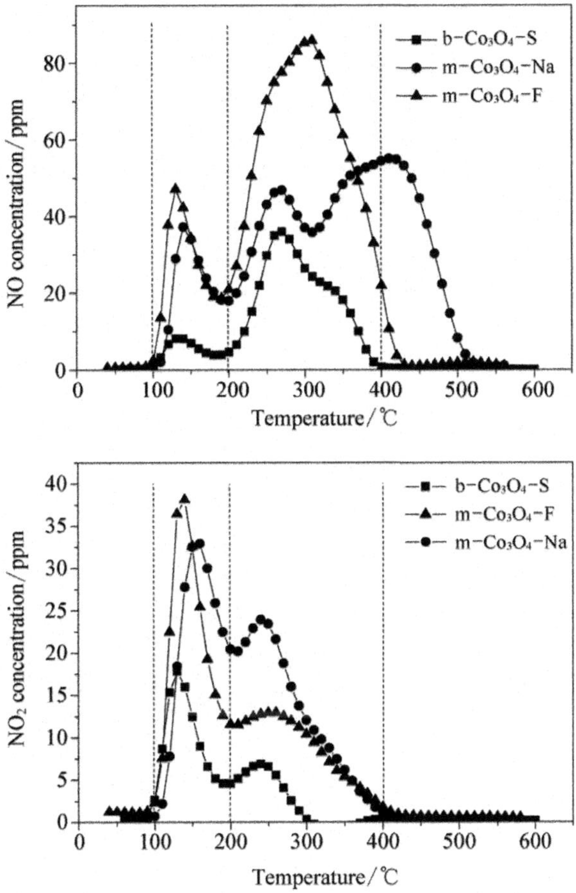

Fig. 2.47 NO and NO_2 concentration through the NO_x–TPD [87]

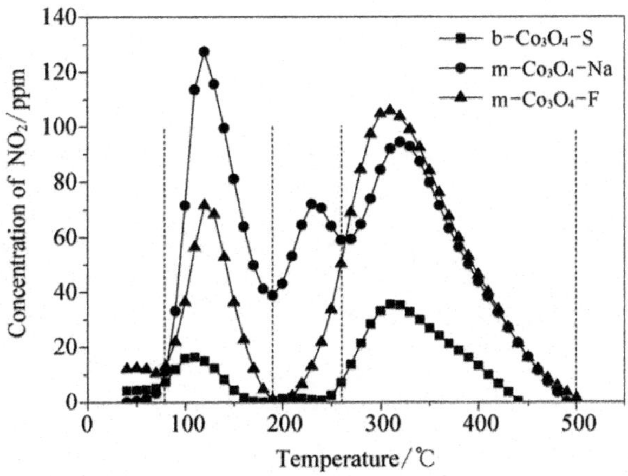

Fig. 2.48 NO_2 concentration of through NO_x–TPO run [87]

shown in NO_x–TPD. It can be attributed to N_2O_4 and/or $NO_2^{\delta+}$ species formed from initial NO adsorption at room temperature and removal as the temperature increased. These initially formed species may be the result of direct interaction between NO and the oxygen species associated with Co_3O_4. Apparently, these species were weakly bonded. The second stage range from 200 to 500 °C with the peak at 310 °C was related to the oxidation of NO and further decomposition of NO_2 with the increase of the temperature [104].

Compared with m–Co_3O_4–F and b–Co_3O_4–S, the m–Co_3O_4–Na sample exhibited different feature during TPO run. It displayed three peaks centered at 120, 230, and 310 °C, respectively. The two peaks centered at 120 and 310 °C for m–Co_3O_4–Na were in the same ranges as m–Co_3O_4–F and b–Co_3O_4–S. The additional shoulder peak centered at 230 °C(range 200–265 °C) with a high amount of NO_2 was consistent with the high-temperature NO_2 desorption peak in NO_x–TPD (Fig. 2.49), which was suggested to arise from the decomposition of nitrates species on the surface of m–Co_3O_4–Na. For m–Co_3O_4–Na, such high amount of NO_2 released between 200 and 265 °C shown in NO–TPO was consistent with the low ignite temperature (250 °C, T_{10}, Table 2.15), which suggested that the formation of nitrates species on the surface of m–Co_3O_4–Na, and its subsequent interaction with soot in the middle-temperature range may be the rate-determining step for simultaneous soot–NO_x reaction over Co_3O_4 catalysts.

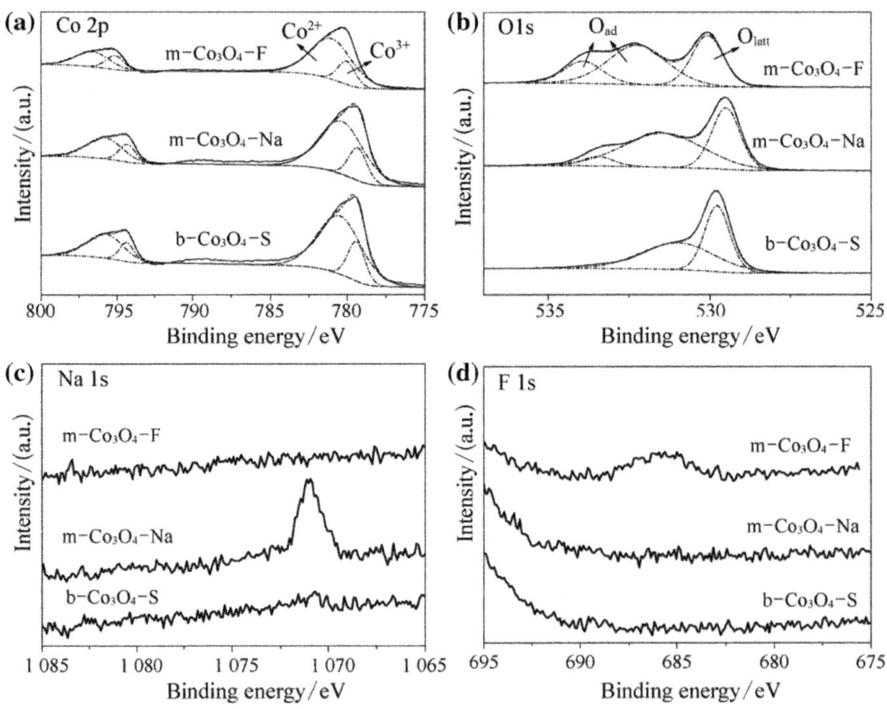

Fig. 2.49 Co 2p (**a**), O 1s (**b**), Na 1s (**c**) and F 1s (**d**) XPS spectra of Co_3O_4 [87]

The above characterization results suggested that nitrates formed on m–Co_3O_4–Na may be the dominant factor in promoting NO_x–soot removal reaction. To clarify whether there was a relationship between the activity and the specific chemical species on the catalyst, XPS and HAADF-STEM were carried out to analyze the surface properties.

The XPS spectra of Co 2p and O 1s on different Co_3O_4 catalysts were displayed in Fig. 2.49a, b. The molar ratio of Co^{3+}/Co^{2+} and $O_{ad}/(O_{latt} + O_{ad})$ was illustrated in Table 2.17. The O 1s after peak fitting procedure showed two main contributions, the O1s peak centered at 529.8 eV assigned to the lattice oxygen in the Co_3O_4 phase (Co–O bond) and the peaks range between 531.4–532.2 and 533.4–533.9 eV on the mesoporous Co_3O_4 indicating the surface-absorbed oxygen species (nonequivalent hydroxyl groups or chemisorbed water) [105–107]. It was found that the ratio of absorbed oxygen to total surface oxygen of m–Co_3O_4 (63 and 66%) was higher than that of b–Co_3O_4–S (54.3%), suggesting the improved redox ability due to the mesoporous structure. However, the ratio of the adsorbed oxygen to the total surface oxygen is just about the same level for two mesoporous samples. The presence of a Co $2p_{3/2}$ peak at 780 eV and a Co $2p_{1/2}$ peak at 795 eV indicated the existence of the Co_3O_4 phase [106–108]. The Co^{3+}/Co^{2+} molar ratio of mesoporous Co_3O_4 was higher than b–Co_3O_4–S, but it showed no significant difference between mesoporous Co_3O_4, which was corresponding to the H_2–TPR and O 1s XPS results.

XPS and ICP results on sodium and fluorine finally illustrated that the significant difference between two mesoporous Co_3O_4 was the surface element residue due to different methods for removing the hard template. Sodium (Na) content and fluorine (F) content were illustrated in Table 2.15. As shown in Fig. 2.49c, XPS spectra for Na 1s peaks were only found on the m–Co_3O_4–Na surface, while it did not exist on other Co_3O_4 samples surface. And F spectra were only found on the m–Co_3O_4–F. The amount of Na was about 0.5 atomic % based on XPS or 1.6 wt% according to ICP (Table 2.17). Since the catalyst was not calcined after removing template with NaOH, the Na species can be assigned to free surface Na species. And the presence of Na 1s peak at 1071 eV indicated the possible existence of the Na_2CO_3 (Na 1s 1071.5 eV) or $NaHCO_3$ (Na 1s 1071.3 eV) on the surface as NaOH (Na 1s 1072.6 eV) might absorb CO_2 in the air and transferred to Na carbonate species during drying process [105]. The presence and homogeneously dispersion of such Na species on the mesoporous framework were further confirmed by the element mapping as shown in Fig. 2.50. As Co_3O_4 could be exposed in the condition that

Table 2.17 Na content, F content, Co^{3+}/Co^{2+} and $O_{ad}/(O_{latt} + O_{ad})$ molar ratio [87]

Sample	Co^{3+}/Co^{2+} molar ratio	$O_{ad}/(O_{latt} + O_{ad})$ molar ratio (%)	Na 1s content (at.%)	F 1s content (at.%)	Na* (wt%)
b–Co_3O_4–S	0.2514	54.3	–	–	–
m–Co_3O_4–Na	0.2608	63.5	0.5	–	1.6
m–Co_3O_4–F	0.2616	66.3	–	0.35	–

*The result of ICP

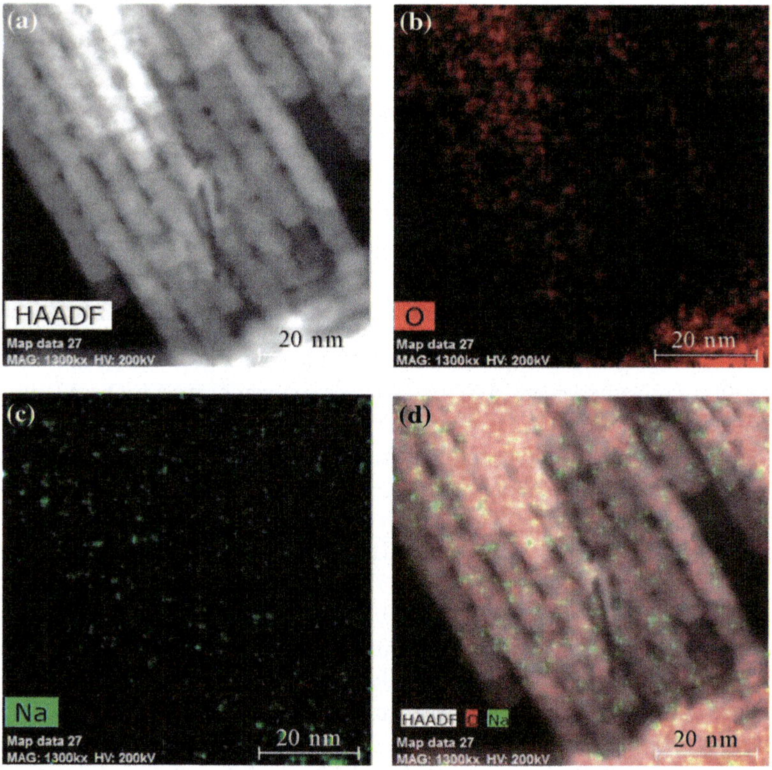

Fig. 2.50 Dark field image (**a**) and element mapping images O (**b**), Na (**c**) and O + Na (**d**) of m–Co_3O_4–Na [87]

the silica was washed away; thus, it is of the high possibility that Na species would disperse not only on the surface but also in the channel structure and interparticle voids.

Note that above results already revealed the different surface functionalized properties due to the remaining of sodium or fluorine species on m–Co_3O_4 using different hard template removal methods, in situ DRIFTS experiments were further carried out to identify the NO_x adsorption species over different Co_3O_4 and reveal the possible relationship between surface functionalization and reaction mechanism. As shown in Fig. 2.51a, b, NO_x adsorbed species were similar for m–Co_3O_4–F and b–Co_3O_4–S. The dominant adsorption species could be ascribed to chelating bidentate nitrate (1548–1560 and 1216–1220 cm^{-1}) and nitrite (1488–1496 and 1271–1288 cm^{-1}) together with adsorbed NO_2 (1619 cm^{-1}) [109–111]. The nitrite species quickly disappeared when the temperature increased above 150 °C, and the bidentate nitrates sharply increased, revealing the formation of stable nitrates with the elevation of temperature. Thus, the mesoporous structure or fluorine residue on the surface of m–Co_3O_4–F does not appear to affect the adsorbed NO_x species significantly. But the relative adsorption amount of NO_x on m–Co_3O_4–F was still

Fig. 2.51 In situ DRIFTS spectra of NO_x sorption behavior at a different temperature over b–Co_3O_4–S (**a**), m–Co_3O_4–F (**b**) and m–Co_3O_4–Na (**c**) [87]

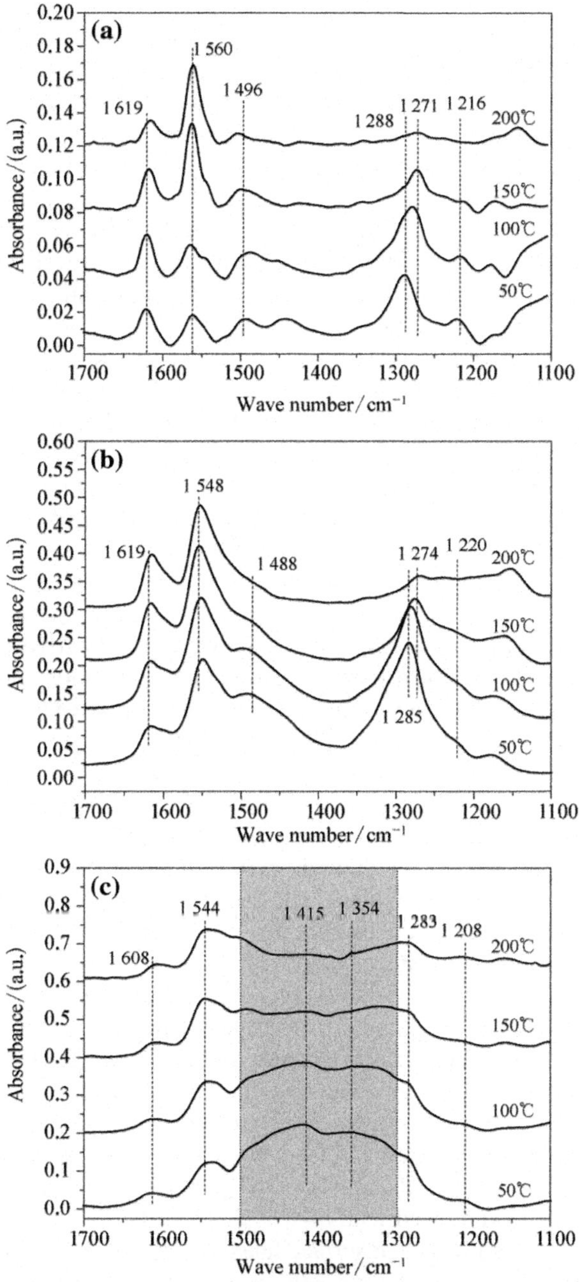

larger than that of b–Co_3O_4–S. Thus, identical conclusions concerning NO_x adsorption amount could be obtained from in situ DRIFTS and NO_x–TPD results, which may help to explain the moderately promotional effect on the activity in NO_x conversion for m–Co_3O_4–F compared with b–Co_3O_4–S.

However, as shown in Fig. 2.51c, it is important to highlight that there is a significant difference on NO_x adsorbed profile between m–Co_3O_4–Na and m–Co_3O_4–F or b–Co_3O_4–S. The appearance of the broad adsorption peak between 1300 and 1500 cm^{-1} (Fig. 2.51c), which could be attributed to ionic nitrates and/or monodentate nitrates [109], was entirely different from the lack of adsorption band in that range as shown in Fig. 2.51a, b, despite the appearance of the band of chelating bidentate nitrate, nitrite, and NO_2 (Fig. 2.51c) at 1544 and 1208, 1283, and 1608 cm^{-1}, respectively. Several different types of nitrates/nitrites appeared on m–Co_3O_4–Na, indicating different active sites were present as m–Co_3O_4–Na was functionalized by sodium. The above results thus provide direct evidence to verify that the introduction of Na facilitates the formation of more kinds of stored N-containing species, corresponding to more substantial NO_x trapping efficiency and higher simultaneously NO_x and soot removal efficiency. It is worth noting that after increasing the temperature from 50 to 200 °C, the band ranging from 1300 to 1500 cm^{-1} becomes weaker strikingly for m–Co_3O_4–Na, compared with other species, arising from the decomposition of ionic nitrate and/or monodentate nitrates. Such behavior could relate to the production of NO_2 in the low temperature as shown in NO_x–TPD, thus possibly promote the soot and NO_x simultaneously removal reaction at relatively lower temperature.

For the NO_x–soot catalysis reaction, various modifiers or additives have already been utilized to improve the activity and selectivity of the catalysts. Shangguan et al. investigated the effect of alkali metal cations doping (Li, Na, K, Cs) for the simultaneous NO_x–soot removal. The doped catalyst was prepared by the citric acid-aided process, and the impregnated catalyst was calcined at 700 °C for 1 h [113]. Potassium (K) displayed activities for the simultaneous removal of nitric oxide and soot which was consistent with others studies [114–116]. Potassium surface modifications were believed to have a number of effects such as modification the electron donor properties of the oxide which may lead to the formation of active forms of oxygen. The potassium cations will also lead to the formation of low melting point compounds or superficial carbonate intermediates which may improve soot–catalyst contact [114, 117, 118].

However, sodium (Na) was reported to improve the activity (decreasing T_{ig}) but deteriorate the conversion of NO_x in a previous study [113]. Thus, potassium is the most popular alkali metal cations modification in the soot combustion or soot–NO_x simultaneously removal reaction. On the contrary, our results illustrated that m–Co_3O_4–Na had shown prominent enhanced simultaneous NO_x and soot removal activity with sodium surface functionalization compared with other Co_3O_4 samples. But m–Co_3O_4–Na also showed negligible enhanced soot oxidation ability under oxygen compared with m–Co_3O_4–F. As HRTEM, XRD, H_2–TPR, and XPS results already showed that surface functionalization has no influence on the crystallization of Co phases, or enhance the release of active Co phases, the different performance of two m–Co_3O_4 in soot–NO–O_2 and soot–O_2 reaction suggested other determinants underlying the evolution profile of NO_x for different Co_3O_4 samples. The NO_x–TPO/TPD provided clues for the existence of nitrate intermediates on the m–Co_3O_4–Na surface which was further verified by the in situ DRIFTS. The decomposition of the unstable nitrate species will release NO_2 in the temperature range 200–265 °C according to the NO_x–TPO/TPD results, corresponding to the ignition temperature range (200–250 °C) for m–Co_3O_4–Na in NO_x–soot removal reaction. The formation of low melting point compounds may also help to enhance the interaction of soot and catalyst with NO_x, thus further promote the soot ignition and NO_x conversion.

Compared with m–Co_3O_4–Na, m–Co_3O_4–F, and b–Co_3O_4–S only showed gas-phase NO oxidation peak in the temperature range 200–265 °C in NO–TPO accompanied with an inferior performance for ignition and selectivity. Thus, the formation of unstable nitrates as supported by the in situ DRIFTS and its further interaction with soot in the middle-temperature range (200–265 °C), i.e., NO reaction pathway, could be the most significant step in the simultaneous soot–NO_x reaction using Co_3O_4 as a catalyst. The formation of nitrates may occur in the following stoichiometry reactions [103, 119, 120]:

$$2NO + 1.5O_2 + O^*_{(s)} \rightarrow 2NO^-_{3(s)} \tag{2.7}$$

$$NO + O_2 \rightarrow NO_2 \tag{2.8}$$

$$NO_2 + 0.5O_2 + O^*_{(s)} \rightarrow 2NO^-_{3(s)} \tag{2.9}$$

These reactions illustrated the evolution of nitrates, involving the oxidation of NO by gaseous or activated oxygen on the catalyst surface. In the case of m–Co_3O_4–Na, the basic component Na may facilitate the adsorption and activation of the oxidative O_2 and NO molecules. NO_x was absorbed and transferred into unstable nitrate intermediates. This nitrate species may play an essential role in soot oxidation for the lower melting temperature, which can strengthen the contact with soot and the less stability to form higher oxidizing species. While for m–Co_3O_4–F, the lack of sodium on the surface leads to the similar NO_x–soot reaction route as b–Co_3O_4–S. Thus, the mesoporous structure of m–Co_3O_4–F only slightly favors the NO reduction by ensuring optimal accessibility gas-phase reactants which lead to higher selectivity to N_2 compared with b–Co_3O_4–S while keeping the ignition performance invariant.

Thus, this study demonstrates that surface chemical species introduced by silica removal procedure together with the mesoporous structure emerge as critical factors regulating specific catalysis reaction. Hard template removal is one specific procedure during nanocasting. Although the overall reaction mechanism and the rate-determining step of specific reaction over a given mesoporous catalyst depend significantly on the nature of the reaction together with the experiment condition, considering that the surface sodium species are so important in this case, one must consider decouple the contributions of factors, such surface chemical species and morphology to specific catalysis reaction during evaluating the impact of mesoporous structure. On the other hand, as alkali metal was already proved to be excellent promoter candidate for various catalysis reactions especially for the soot and NO_x simultaneously removal reaction mentioned in this book [13, 116, 121, 122], removal silica using hot NaOH solution thus provide a facile route to achieve homogeneous dispersion of sodium species not only in the surface but also the whole framework of mesoporous materials.

References

1. K. Yoshida, S. Makino, S. Sumiya, G. Muramatsu, R. Helferich, Simultaneous reduction of NO_x and particulate emissions from diesel engine exhaust. SAE Technical Paper (1989)
2. O. Kralov, *Catalysis by Non Metals* (Academic Press, New York, 1970)
3. J. Choa, M.M. Thackeray, Structural changes of $LiMn_2O_4$ spinel electrodes during electrochemical cycling. J. Electrochem. Soc. **146**, 3577–3581 (1999)
4. N. Sahli, C. Petit, A.-C. Roger, A. Kiennemann, S. Libs, M.M. Bettahar, Ni catalysts from $NiAl_2O_4$ spinel for CO_2 reforming of methane. Catal. Today **113**, 187–193 (2006)
5. W. Shangguan, Y. Teraoka, S. Kagawa, Simultaneous catalytic removal of NO_x and diesel soot particulates over ternary AB_2O_4 spinel-type oxides. Appl. Catal. B **8**, 217–227 (1996)
6. H.H. Kung, *Transition Metal Oxides: Surface Chemistry and Catalysis* (Elsevier, 1989)
7. A.Y. Khodakov, Fischer-Tropsch synthesis: relations between structure of cobalt catalysts and their catalytic performance. Catal. Today **144**, 251–257 (2009)
8. F. Kapteijn, AJC Mierop. G. Abbel. and JA Moulijn. Chem. Soc. Chem. Commun. 1085 (1984)
9. A. Ahlstrom, I. Odenbrand, Combustion of soot deposits from diesel engines on mixed oxides of vanadium pentoxide and cupric oxide. Appl. Catal. **60**, 157–172 (1990)
10. A.F. Ahlström, C.I. Odenbrand, Catalytic combustion of soot deposits from diesel engines. Appl. Catal. **60**, 143–156 (1990)
11. P. Ciambelli, P. Corbo, M. Gambino, V. Palma, S. Vaccaro, Catalytic combustion of carbon particulate. Catal. Today **27**, 99–106 (1996)
12. C. Badini, V. Serra, G. Saracco, M. Montorsi, Thermal stability of Cu-KV catalyst for diesel soot combustion. Catal. Lett. **37**, 247–254 (1996)
13. W. Shangguan, Y. Teraoka, S. Kagawa, Promotion effect of potassium on the catalytic property of $CuFe_2O_4$ for the simultaneous removal of NO_x and diesel soot particulate. Appl. Catal. B **16**, 149–154 (1998)
14. Y. Teraoka, K. Nakano, W. Shangguan, S. Kagawa, Simultaneous catalytic removal of nitrogen oxides and diesel soot particulate over perovskite-related oxides. Catal. Today **27**, 107–113 (1996)
15. M.C. Biesinger, L.W. Lau, A.R. Gerson, R.S.C. Smart, Resolving surface chemical states in XPS analysis of first row transition metals, oxides and hydroxides: Sc, Ti, V, Cu and Zn. Appl. Surf. Sci. **257**, 887–898 (2010)
16. T. Yamashita, P. Hayes, Analysis of XPS spectra of Fe^{2+} and Fe^{3+} ions in oxide materials. Appl. Surf. Sci. **254**, 2441–2449 (2008)
17. W. Shangguan, Y. Teraoka, S. Kagawa, Effect of oxide composition of spinel-type copper chromites on the catalytic activity for the simultaneous removal of NO_x and soot particulate. Res. Rep. Dept. Eng. Nagasaki Univ. **25**, 241–248 (1995)
18. Y. Teraoka, K. Nakano, S. Kagawa, W. Shangguan, Simultaneous removal of nitrogen oxides and diesel soot particulates catalyzed by perovskite-type oxides. Appl. Catal. B **5**, L181–L185 (1995)
19. R. Voorhoeve, D. Johnson, J. Remeika, P. Gallagher, Perovskite oxides: materials science in catalysis. Science **195**, 827–833 (1977)
20. L.G. Tejuca, J.L.G. Fierro, J.M. Tascón, Structure and reactivity of perovskite-type oxides. Adv. Catal. **36**, 237–328 (1989)

21. C.H. Kim, G. Qi, K. Dahlberg, W. Li, Strontium-doped perovskites rival platinum catalysts for treating NO_x in simulated diesel exhaust. Science **327**, 1624–1627 (2010)
22. P. Ciambelli, S. Cimino, S. De Rossi, L. Lisi, G. Minelli, P. Porta, G. Russo, AFeO 3 (A = La, Nd, Sm) and $LaFe_{1-x}Mg_xO_3$ perovskites as methane combustion and CO oxidation catalysts: structural, redox and catalytic properties. Appl. Catal. B **29**, 239–250 (2001)
23. S. Cimino, S. Colonna, S. De Rossi, M. Faticanti, L. Lisi, I. Pettiti, P. Porta, Methane combustion and CO oxidation on zirconia-supported La, Mn oxides and $LaMnO_3$ perovskite. J. Catal. **205**, 309–317 (2002)
24. N. Yamazoe, Y. Teraoka, Oxidation catalysis of perovskites—relationships to bulk structure and composition (valency, defect, etc.). Catal. Today **8**, 175–199 (1990)
25. C. Ramana, M. Massot, C. Julien, XPS and Raman spectroscopic characterization of $LiMn_2O_4$ spinels. Surf. Interface Anal. **37**, 412–416 (2005)
26. M.A. Stranick, MnO_2 by XPS. Surf. Sci. Spectra **6**, 31–38 (1999)
27. M.A. Stranick, Mn_2O_3 by XPS. Surf. Sci. Spectra **6**, 39–46 (1999)
28. K. Tabata, Y. Hirano, E. Suzuki, XPS studies on the oxygen species of $LaMn_{1-x}Cu_xO_{3+\lambda}$. Appl. Catal. A **170**, 245–254 (1998)
29. X. Peng, H. Lin, Z. Huang, W. Shangguan, Influence of composition of perovskite-type catalysts $La_{(1-x)}A_xB_yMn_{(1-y)}O_3$ on simultaneous removal of NO_x and soot. J. Chem. Eng. Chin. Univ. **5**, 028 (2006)
30. Y. Teraoka, K. Kanada, S. Kagawa, Synthesis of La-K-Mn-O perovskite-type oxides and their catalytic property for simultaneous removal of NO_x and diesel soot particulates. Appl. Catal. B **34**, 73–78 (2001)
31. X. Peng, H. Lin, W. Shangguan, Z. Huang, Physicochemical and catalytic properties of $La_{0.8}K_{0.2}Cu_xMn_{1-x}O_3$ for simultaneous removal of NO_x and soot: effect of Cu substitution amount and calcination temperature. Ind. Eng. Chem. Res. **45**, 8822–8828 (2006)
32. M. Misono, A view on the future of mixed oxide catalysts: the case of heteropolyacids (polyoxometalates) and perovskites. Catal. Today **100**, 95–100 (2005)
33. X. Peng, H. Lin, W. Shangguan, Z. Huang, The effect of partial substitution K and Cu in $LaMnO_3$ perovskites for simultaneous soot-NO_x removal. Chin. J. Catal. **27**, 767–771 (2006). (in Chinese)
34. X. Peng, H. Lin, W. Shangguan, Z. Huang, Surface properties and catalytic performance of $La_{0.8}K_{0.2}Cu_xMn_{1-x}O_3$ for simultaneous removal of NO_x and soot. Chem. Eng. Technol. **30**, 99–104 (2007)
35. D.G. Evans, R.C. Slade, Structural aspects of layered double hydroxides, in *Layered Double Hydroxides* (Springer, 2006), pp. 1–87
36. A. Vaccari, Preparation and catalytic properties of cationic and anionic clays. Catal. Today **41**, 53–71 (1998)
37. L. Chmielarz, P. Kuśtrowski, A. Rafalska-Łasocha, D. Majda, R. Dziembaj, Catalytic activity of Co-Mg-Al, Cu-Mg-Al and Cu-Co-Mg-Al mixed oxides derived from hydrotalcites in SCR of NO with ammonia. Appl. Catal. B: Environ. **35**, 195–210 (2002)
38. A. Corma, A. Palomares, F. Rey, F. Márquez, Simultaneous catalytic removal of SO_x and NO x with hydrotalcite-derived mixed oxides containing copper, and their possibilities to be used in FCC units. J. Catal. **170**, 140–149 (1997)
39. S. Kannan, Decomposition of nitrous oxide over the catalysts derived from hydrotalcite-like compounds. Appl. Clay Sci. **13**, 347–362 (1998)
40. L. Chmielarz, P. Kuśtrowski, M. Zbroja, A. Rafalska-Łasocha, B. Dudek, R. Dziembaj, SCR of NO by NH_3 on alumina or titania-pillared montmorillonite various modified with Cu or Co: part I. General characterization and catalysts screening. Appl. Catal. B: Environ. **45**, 103–116 (2003)
41. S. Velu, K. Suzuki, T. Osaki, A comparative study of reactions of methanol over catalysts derived from NiAl-and CoAl-layered double hydroxides and their Sn-containing analogues. Catal. Lett. **69**, 43–50 (2000)
42. P.G. Harrison, I.K. Ball, W. Daniell, P. Lukinskas, M.A. Céspedes, E.E. Miró, M.A.A. Ulla, Cobalt catalysts for the oxidation of diesel soot particulate. Chem. Eng. J. **95**, 47–55 (2003)

43. C. Querini, M. Ulla, F. Requejo, J. Soria, U. Sedrán, E. Miró, Catalytic combustion of diesel soot particles. Activity and characterization of Co/MgO and Co, K/MgO catalysts. Appl. Catal. B: Environ. **15**, 5–19 (1998)
44. S. Kureti, W. Weisweiler, K. Hizbullah, Simultaneous conversion of nitrogen oxides and soot into nitrogen and carbon dioxide over iron containing oxide catalysts in diesel exhaust gas. Appl. Catal. B **43**, 281–291 (2003)
45. E. Miró, F. Ravelli, M. Ulla, L. Cornaglia, C. Querini, Catalytic combustion of diesel soot on Co, K supported catalysts. Catal. Today **53**, 631–638 (1999)
46. Z. Wang, Z. Jiang, W. Shangguan, Simultaneous catalytic removal of NO_x and soot particulate over Co–Al mixed oxide catalysts derived from hydrotalcites. Catal. Commun. **8**, 1659–1664 (2007)
47. Z. Wang, W. Shangguan, J. Su, Z. Jiang, Catalytic oxidation of diesel soot on mixed oxides derived from hydrotalcites. Catal. Lett. **112**(3–4), 149–154 (2006)
48. C. Querini, L. Cornaglia, M. Ulla, E. Miro, Catalytic combustion of diesel soot on Co, K/ MgO catalysts. Effect of the potassium loading on activity and stability. Appl. Catal. B: Environ. **20**, 165–177 (1999)
49. W. Shangguan, Y. Teraoka, S. Kagawa, Kinetics of soot—O_2, soot—NO and soot—O_2—NO reactions over spinel-type $CuFe_2O_4$ catalyst. Appl. Catal. B **12**, 237–247 (1997)
50. V. Milt, M. Pissarello, E. Miró, C. Querini, Abatement of diesel-exhaust pollutants: NO_x storage and soot combustion on K/La_2O_3 catalysts. Appl. Catal. B **41**, 397–414 (2003)
51. M. Machida, Y. Murata, K. Kishikawa, D. Zhang, K. Ikeue, On the reasons for high activity of CeO_2 catalyst for soot oxidation. Chem. Mater. **20**, 4489–4494 (2008)
52. S. Aouad, E. Saab, E. Abi-Aad, A. Aboukaïs, Study of the Ru/Ce system in the oxidation of carbon black and volatile organic compounds. Kinet. Catal. **48**, 835–840 (2007)
53. Z. Li, M. Meng, Y. Zha, F. Dai, T. Hu, Y. Xie, J. Zhang, Highly efficient multifunctional dually-substituted perovskite catalysts $La_{1-x}K_xCo_{1-y}Cu_yO_{3-\delta}$ used for soot combustion, NO_x storage and simultaneous NO_x-soot removal. Appl. Catal. B: Environ. **121**, 65–74 (2012)
54. S.-S. Hong, G.-D. Lee, Simultaneous removal of NO and carbon particulates over lanthanoid perovskite-type catalysts. Catal. Today **63**, 397–404 (2000)
55. G. Zou, M. Chen, W. Shangguan, Promotion effects of $LaCoO_3$ formation on the catalytic performance of Co–La oxides for soot combustion in air. Catal. Commun. **51**, 68–71 (2014)
56. E. Makshina, S. Sirotin, V. Yushchenko, G. Mazo, M. van den Berg, K. Klements'ev, W. Grünert, B. Romanovskii, Nanocomposites based on $LaCoO_3$ and mesoporous molecular sieves: preparation and physicochemical and catalytic properties. Kinet. Catal. **47**, 49–53 (2006)
57. L.F. Liotta, M. Ousmane, G. Di Carlo, G. Pantaleo, G. Deganello, A. Boreave, A. Giroir-Fendler, Catalytic removal of toluene over Co_3O_4–CeO_2 mixed oxide catalysts: comparison with Pt/Al_2O_3. Catal. Lett. **127**, 270–276 (2008)
58. J. Ren, Y. Yu, F. Dai, M. Meng, J. Zhang, L. Zheng, T. Hu, Domain-confined catalytic soot combustion over Co_3O_4 anchored on a TiO_2 nanotube array catalyst prepared by mercaptoacetic acid induced surface-grafting. Nanoscale **5**, 12144–12149 (2013)
59. H. Wang, Z. Zhao, P. Liang, C. Xu, A. Duan, G. Jiang, J. Xu, J. Liu, Highly Active $La_{1-x}K_x$ CoO_3 perovskite-type complex oxide catalysts for the simultaneous removal of diesel soot and nitrogen oxides under loose contact conditions. Catal. Lett. **124**, 91–99 (2008)
60. Z. Li, M. Meng, Y. Zha, F. Dai, T. Hu, Y. Xie, J. Zhang, Highly efficient multifunctional dually-substituted perovskite catalysts $La_{1-x}KxCo_{1-y}CuyO_{3-\delta}$ used for soot combustion, NO_x storage and simultaneous NOx-soot removal. Appl. Catal. B Environ. **121**, 65–74 (2012)
61. P. Kirienko, S. Solov'ev, S. Orlik, Effect of CeO_2 on the properties of the Pd/Co3O4/ cordierite catalyst in the conversion of CO, NO, and hydrocarbons. Theor. Exp. Chem. **46**, 39–44 (2010)
62. G. Zou, Y. Xu, S. Wang, M. Chen, W. Shangguan, The synergistic effect in Co–Ce oxides for catalytic oxidation of diesel soot. Catal. Sci. Technol. **5**, 1084–1092 (2015)

63. V. Hadjiev, M. Iliev, I. Vergilov, The Raman spectra of Co_3O_4. J. Phys. C: Solid State Phys. **21**, L199 (1988)
64. I. Lorite, J.J. Romero, J.F. Fernández, Effects of the agglomeration state on the Raman properties of Co_3O_4 nanoparticles. J. Raman Spectrosc. **43**, 1443–1448 (2012)
65. W.S. Kijlstra, D.S. Brands, E.K. Poels, A. Bliek, Mechanism of the selective catalytic reduction of NO by NH_3 over MnO_x/Al_2O_3. J. Catal. **171**, 208–218 (1997)
66. C. Vaz, H.-Q. Wang, C. Ahn, V. Henrich, M. Baykara, T. Schwendemann, N. Pilet, B. Albers, U. Schwarz, L. Zhang, Interface and electronic characterization of thin epitaxial Co_3O_4 films. Surf. Sci. **603**, 291–297 (2009)
67. J. Liu, Z. Zhao, J. Wang, C. Xu, A. Duan, G. Jiang, Q. Yang, The highly active catalysts of nanometric CeO_2-supported cobalt oxides for soot combustion. Appl. Catal. B **84**, 185–195 (2008)
68. M.M. Natile, A. Glisenti, CoO_x/CeO_2 nanocomposite powders: synthesis, characterization, and reactivity. Chem. Mater. **17**, 3403–3414 (2005)
69. M. O'Connell, A. Norman, C. Hüttermann, M. Morris, Catalytic oxidation over lanthanum-transition metal perovskite materials. Catal. Today **47**, 123–132 (1999)
70. S. Shinde, K. Rajpure, X-ray photoelectron spectroscopic study of catalyst based zinc oxide thin films. J. Alloy. Compd. **509**, 4603–4607 (2011)
71. E. Bêche, P. Charvin, D. Perarnau, S. Abanades, G. Flamant, Ce 3d XPS investigation of cerium oxides and mixed cerium oxide ($Ce_xTi_yO_z$). Surf. Interface Anal. **40**, 264–267 (2008)
72. B.M. Reddy, A. Khan, Y. Yamada, T. Kobayashi, S. Loridant, J.-C. Volta, Structural characterization of CeO_2-TiO_2 and V_2O_5/CeO_2-TiO_2 catalysts by Raman and XPS techniques. J. Phys. Chem. B **107**, 5162–5167 (2003)
73. H.B. Pereira, C. Polato, J.L.F. Monteiro, C.A. Henriques, Mn/Mg/Al-spinels as catalysts for SOx abatement: influence of CeO_2 incorporation and catalytic stability. Catal. Today **149**, 309–315 (2010)
74. G. Colón, J. Navio, R. Monaci, I. Ferino, CeO_2–La_2O_3 catalytic system Part I. Preparation and characterisation of catalysts. Phys. Chem. Chem. Phys. **2**, 4453–4459 (2000)
75. D. Zhao, J. Feng, Q. Huo, N. Melosh, G.H. Fredrickson, B.F. Chmelka, G.D. Stucky, Triblock copolymer syntheses of mesoporous silica with periodic 50 to 300 angstrom pores. Science **279**, 548–552 (1998)
76. Y. Ren, Z. Ma, P.G. Bruce, Ordered mesoporous metal oxides: synthesis and applications. Chem. Soc. Rev. **41**, 4909–4927 (2012)
77. A. Taguchi, F. Schüth, Ordered mesoporous materials in catalysis. Microporous Mesoporous Mater. **77**, 1–45 (2005)
78. C.M.A. Parlett, K. Wilson, A.F. Lee, Hierarchical porous materials: catalytic applications. Chem. Soc. Rev. **42**, 3876–3893 (2013)
79. C.M. Parlett, M.A. Isaacs, S.K. Beaumont, L.M. Bingham, N.S. Hondow, K. Wilson, A.F. Lee, Spatially orthogonal chemical functionalization of a hierarchical pore network for catalytic cascade reactions. Nat. Mater. **15**, 178–182 (2016)
80. C. Perego, R. Millini, Porous materials in catalysis: challenges for mesoporous materials. Chem. Soc. Rev. **42**, 3956–3976 (2013)
81. Y. Wei, Z. Zhao, J. Liu, C. Xu, G. Jiang, A. Duan, Design and synthesis of 3D ordered macroporous CeO_2–supported Pt@$CeO_{2-\delta}$ core-shell nanoparticle materials for enhanced catalytic activity of soot oxidation. Small **9**, 3957–3963 (2013)
82. Y. Wang, H. Arandiyan, J. Scott, A. Bagheri, H. Dai, R. Amal, Recent advances in ordered meso/macroporous metal oxides for heterogeneous catalysis: a review. J. Mater. Chem. A **5**, 8825–8846 (2017)
83. S. Van Donk, A.H. Janssen, J.H. Bitter, K.P. de Jong, Generation, characterization, and impact of mesopores in zeolite catalysts. Catal. Rev. **45**, 297–319 (2003)
84. Q. Ding, H. Xian, Y. Tan, N. Tsubaki, X. Li, Mesoporous SiO_2-confined $La_{0.7}Sr_{0.3}CoO_3$ perovskite nanoparticles: an efficient NO_x adsorber for lean-burn exhausts. Catal. Sci. Technol. **3**, 1493–1496 (2013)

85. Y. Du, Q. Meng, J. Wang, J. Yan, H. Fan, Y. Liu, H. Dai, Three-dimensional mesoporous manganese oxides and cobalt oxides: high-efficiency catalysts for the removal of toluene and carbon monoxide. Microporous Mesoporous Mater. **162**, 199–206 (2012)
86. Z. Jiang, W. Zhang, W. Shangguan, X. Wu, Y. Teraoka, Adsorption of NO molecule on spinel-type $CuFe_2O_4$ surface: a first-principles study. J. Phys. Chem. C **115**, 13035–13040 (2011)
87. Z. Jiang, Z. Zhu, W. Guo, M. Chen, W. Shangguan, Surface sodium functionalization of ordered mesoporous Co_3O_4 controls the enhanced simultaneous catalytic removal of soot and NO_x. J. Mater. Chem. A **5**, 20696–20708 (2017)
88. M. Kruk, M. Jaroniec, C.H. Ko, R. Ryoo, Characterization of the porous structure of SBA-15. Chem. Mater. **12**, 1961–1968 (2000)
89. Y. Feng, L. Li, S. Niu, Y. Qu, Q. Zhang, Y. Li, W. Zhao, H. Li, J. Shi, Controlled synthesis of highly active mesoporous Co_3O_4 polycrystals for low temperature CO oxidation. Appl. Catal. B **111**, 461–466 (2012)
90. M. Casascabanas, G. Binotto, D. Larcher, A. Lecup, V. Giordani, J. Tarascon, Defect chemistry and catalytic activity of nanosized Co_3O_4. Chem. Mater. **21**, 1939–1947 (2009)
91. N. Najmoddin, A. Beitollahi, M. Muhammed, N. Ansari, E. Devlin, S.M. Mohseni, H. Rezaie, D. Niarchos, J. Åkerman, M.S. Toprak, Effect of nanoconfinement on the formation, structural transition and magnetic behavior of mesoporous copper ferrite. J. Alloy. Compd. **598**, 191–197 (2014)
92. F. Jiao, A. Harrison, J.-C. Jumas, A.V. Chadwick, W. Kockelmann, P.G. Bruce, Ordered mesoporous Fe_2O_3 with crystalline walls. J. Am. Chem. Soc. **128**, 5468–5474 (2006)
93. B. Bai, H. Arandiyan, J. Li, Comparison of the performance for oxidation of formaldehyde on nano-Co_3O_4, 2D-Co_3O_4, and 3D-Co_3O_4 catalysts. Appl. Catal. B **142**, 677–683 (2013)
94. X. Xie, Y. Li, Z.-Q. Liu, M. Haruta, W. Shen, Low-temperature oxidation of CO catalysed by Co_3O_4 nanorods. Nature **458**, 746–749 (2009)
95. C.Y. Ma, Z. Mu, J.J. Li, Y.G. Jin, J. Cheng, G.Q. Lu, Z.P. Hao, S.Z. Qiao, Mesoporous Co_3O_4 and Au/Co_3O_4 catalysts for low-temperature oxidation of trace ethylene. J. Am. Chem. Soc. **132**, 2608–2613 (2010)
96. B. Bai, H. Arandiyan, J. Li, Comparison of the performance for oxidation of formaldehyde on nano-Co_3O_4, 2D-Co_3O_4, and 3D-Co_3O_4 catalysts. Appl. Catal. B **142–143**, 677–683 (2013)
97. C.-I. Ahn, D.-W. Jeong, J.M. Cho, H.-S. Na, W.-J. Jang, H.-S. Roh, J.-H. Choi, S.H. Um, J.W. Bae, Water gas shift reaction on the Mn-modified ordered mesoporous Co_3O_4. Microporous Mesoporous Mater. **221**, 204–211 (2016)
98. L. Xue, C. Zhang, H. He, Y. Teraoka, Catalytic decomposition of N_2O over CeO_2 promoted Co_3O_4 spinel catalyst. Appl. Catal. B: Environ. **75**, 167–174 (2007)
99. Y. Peng, W. Si, J. Luo, W. Su, H. Chang, J. Li, J. Hao, J.C. Crittenden, Surface tuning of $La_{0.5}Sr_{0.5}CoO_3$ perovskite catalysts by acetic acid for NO_x storage and reduction. Environ. Sci. Technol. **50**, 6442–6448 (2016)
100. R. Vijay, R.J. Hendershot, S.M. Riverajimenez, W.B. Rogers, B.J. Feist, C.M. Snively, J. Lauterbach, Noble metal free NO_x storage catalysts using cobalt discovered via high-throughput experimentation. Catal. Commun. **6**, 167–171 (2005)
101. L. Sivachandiran, F. Thevenet, P. Gravejat, A. Rousseau, Investigation of NO and NO_2 adsorption mechanisms on TiO_2 at room temperature. Appl. Catal. B: Environ. **142**, 196–204 (2013)
102. J. Szanyi, J.H. Kwak, D.H. Kim, S.D. Burton, C.H.F. Peden, NO_2 adsorption on BaO/Al_2O_3: the nature of nitrate species. J. Phys. Chem. B **109**, 27–29 (2005)
103. F. Prinetto, G. Ghiotti, I. Nova, L. Lietti, E. Tronconi, P. Forzatti, FT-IR and TPD investigation of the NO_x storage properties of BaO/Al_2O_3 and Pt–BaO/Al_2O_3 catalysts. J. Phys. Chem. B **105**, 12732–12745 (2001)
104. L. Olsson, H. Persson, E. Fridell, M. Skoglundh, B. Andersson, A kinetic study of NO oxidation and NO_x storage on Pt/Al_2O_3 and Pt/BaO/Al_2O_3. J. Phys. Chem. B **105**, 6895–6906 (2001)

105. C.D. Wagner, G.E. Muilenberg, Handbook of X-ray photoelectron spectroscopy: a reference book of standard data for use in X-ray photoelectron spectroscopy, in *Physical Electronics Division* (Perkin-Elmer Corp., 1979)
106. X. Wang, W. Wen, J. Mi, X. Li, R. Wang, The ordered mesoporous transition metal oxides for selective catalytic reduction of NO_x at low temperature. Appl. Catal. B: Environ. **176**, 454–463 (2015)
107. M.M. Natile, A. Glisenti, Study of surface reactivity of cobalt oxides: interaction with methanol. Chem. Mater. **14**, 3090–3099 (2002)
108. C.-I. Ahn, J.W. Bae, Fischer-Tropsch synthesis on the Al_2O_3-modified ordered mesoporous Co_3O_4 with an enhanced catalytic activity and stability. Catal. Today **265**, 27–35 (2016)
109. K.I. Hadjiivanov, Identification of neutral and charged N_xO_y surface species by IR spectroscopy. Catal. Rev. **42**, 71–144 (2000)
110. Y. Shi, S. Chen, H. Sun, Y. Shu, X. Quan, Low-temperature selective catalytic reduction of NO_x with NH_3 over hierarchically macro-mesoporous Mn/TiO_2. Catal. Commun. **42**, 10–13 (2013)
111. C. He, K. Köhler, Selective catalytic reduction of NO by propane over CoO_x/Al_2O_3: an investigation of the surface reactions using in situ infrared spectroscopy. Phys. Chem. Chem. Phys. **8**, 898–905 (2006)
112. M.L. Traulsen, H.H. Ingelsten, K.K. Hansen, Diffuse reflectance infrared fourier transform study of NO_x adsorption on CGO10 impregnated with K_2O or BaO. J. Phys. Chem. A **116**, 2497–2505 (2012)
113. W. Shangguan, Y. Teraoka, S. Kagawa, Promotion effect of potassium on the catalytic property of $CuFe_2O_4$ for the simultaneous removal of NO_x and diesel soot particulate. Appl. Catal. B: Environ. **16**, 149–154 (1998)
114. M.E. Gálvez, S. Ascaso, R. Moliner, M.J. Lázaro, Influence of the alkali promoter on the activity and stability of transition metal (Cu Co, Fe) based structured catalysts for the simultaneous removal of soot and NO_x. Top. Catal. **56**, 493–498 (2013)
115. M.L. Pisarello, V. Milt, M.A. Peralta, C.A. Querini, E.E. Miró, Simultaneous removal of soot and nitrogen oxides from diesel engine exhausts. Catal. Today **75**, 465–470 (2002)
116. E. Aneggi, C. De Leitenburg, G. Dolcetti, A. Trovarelli, Diesel soot combustion activity of ceria promoted with alkali metals. Catal. Today **136**, 3–10 (2008)
117. T. Jakubek, W. Kaspera, P. Legutko, P. Stelmachowski, A. Kotarba, How to efficiently promote transition metal oxides by alkali towards catalytic soot oxidation. Top. Catal. **59**, 1083–1089 (2016)
118. A.M. Hernandezgimenez, D.L. Castello, A. Buenolopez, Diesel soot combustion catalysts: review of active phases. Chem. Pap. **68**, 1154–1168 (2014)
119. T. Baidya, P. Bera, B.D. Mukri, S.K. Parida, O. Krocher, M. Elsener, M.S. Hegde, DRIFTS studies on CO and NO adsorption and NO + CO reaction over Pd^{2+}-substituted CeO_2 and $Ce_{0.75}Sn_{0.25}O_2$ catalysts. J. Catal. **303**, 117–129 (2013)
120. L. Wu, S. Tong, M. Ge, Heterogeneous reaction of NO_2 on Al_2O_3: the effect of temperature on the nitrite and nitrate formation. J. Phys. Chem. A **117**, 4937–4944 (2013)
121. C.S. MacLeod, A.P. Harvey, A.F. Lee, K. Wilson, Evaluation of the activity and stability of alkali-doped metal oxide catalysts for application to an intensified method of biodiesel production. Chem. Eng. J. **135**, 63–70 (2008)
122. R.S. París, V. Montes, M. Boutonnet, S. Järås, Higher alcohol synthesis over nickel-modified alkali-doped molybdenum sulfide catalysts prepared by conventional coprecipitation and coprecipitation in microemulsions. Catal. Today **258**, 294–303 (2015)

Chapter 3
Kinetics Study for Simultaneous Removal of Soot and NO_x

Chemical kinetics is the study of the rate and mechanism by which a chemical species is converted to another. The gas reaction over solid catalysts have been widely explored and investigated; some mechanisms were reliably clear for a specific system. Such catalysts depend on their activity in part, at least, on the extent of surface area because it adsorbs gas reactants and serves as the site where the catalytic reactions proceed [1]. It is especially difficult to resolve the reaction mechanism involving solid since the process itself is unsteady. It is confirmed that the reaction takes place at the solid–solid–gas interface, the specific surface area is of less importance than in the usual solid–gas reaction. This indicates the untraditional and complicated mechanism for this heterogeneous reaction.

3.1 Kinetic Characteristics of Simultaneous NO_x–Soot Removal

3.1.1 Arrhenius Plots and Compensation Effects

The Arrhenius plot of reaction profiles over $CuFe_2O_4$ (Fig. 2.2) is depicted in Fig. 3.1. The linear tendencies are observed below 340 °C, and the reaction rates seem to level off at this point that the amount of soot consumption achieves 24% of initially charged soot. These facts all provide the possibility for kinetic analysis for soot combustion at a lower temperature (340 °C). It is also assumed that the effective area of the soot/catalyst contact stabilizes as a constant. On the basis of the aforementioned analysis, the reaction rate can be written by the conventional power-law expression as follows:

$$\ln r = -E_a/RT + \ln A + n_1 \ln C_C + n_2 \ln NO + n_3 \ln O_2 \quad (3.1)$$

W. Shangguan et al., *Simultaneous Catalytic Removal of Diesel Soot and NO_x*, Energy and Environment Research in China,
https://doi.org/10.1007/978-981-10-7266-6_3

Fig. 3.1 Rates of CO_2, N_2, and N_2O formation over $CuFe_2O_4$ as a function of reciprocal of temperature [2]

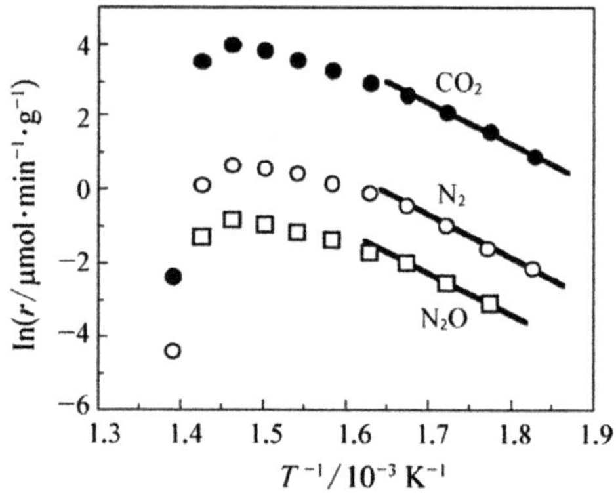

where A is the pre-exponential factor, E_a the apparent energy, C_i the concentration, and n_i the reaction order. The linear relation between ln r and $T - 1$ suggest that Eq. (3.1) is simplified into Eq. (3.2) and that E_a and ln ΓA can be obtained from the slope and the intercept, respectively.

$$\ln r = -E_a/RT + \ln \Gamma A \Gamma = C_C^{n_1} C_{NO}^{n_2} C_{O_2}^{n_3} \tag{3.2}$$

The apparent activation energies (E_a) for the formation of CO_2, N_2, and N_2O calculated from the linear portion of Arrhenius plots are aggregated in Table 3.1. E_a values for CO_2 and N_2 formation are broadly close each other, while that for N_2O formation tends to show a more significant distinction. The minor gap between E_a values for N_2 and CO_2 formation indicates that the selectivity to NO_x reduction into N_2 is temperature independent which is corroborated by the comparisons between rN_2/rCO_2 and $V[N_2]/V[CO_2]$.

A linear correlation or the compensation effect is obtained between E_a and ln ΓA the spinel-type oxides for the formation of CO_2, N_2, and N_2O. This implies that the reaction mechanisms and reaction orders with respect to reactants are the same overall the spinel catalysts studied. It is noteworthy that if the rate NO_x reduction

Table 3.1 Apparent activation energies and selectivity parameters of spinel-type oxides for simultaneous NO_x–soot removal

Catalyst	E_a (kJ mol^{-1})			$S_{N2/C}$	
	CO_2	N_2	N_2O	r_{N2}/r_{CO}	$V[N_2]/V[N_2O]$
$CuCr_2O_4$	86.0	83.7	104.4	2.8	2.6
$MgCr_2O_4$	89.2	82.5	94.6	2.8	2.7
$CoCr_2O_4$	87.0	82.0	119.2	1.7	1.8
$MnCr_2O_4$	82.9	76.3	94.6	1.6	1.5
$CoMn_2O_4$	81.5	77.7	101.1	1.8	1.9
$NiFe_2O_4$	77.5	85.9	80.8	3.1	3.6
$CoFe_2O_4$	95.4	80.1	129.9	1.7	1.8
$CuFe_2O_4$	90.6	91	89.1	4.6	4.7

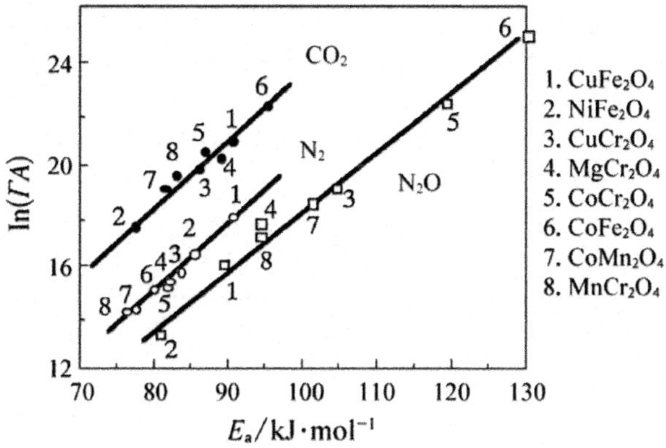

Fig. 3.2 Compensation effects with respect to CO_2, N_2, and N_2O formation in simultaneous NO_x–soot removal over spinel-type oxides [2]

into N_2 and N_2O ($r[N_2] + r[N_2O]$) are applied, E_a for CO_2 formation and NO_x reduction is identical to each other; this also demonstrated that the redox process occurs between CO_2 and NO_x rather than O_2 (Fig. 3.2).

Furthermore, the linear relation between E_a for N_2 formation ($E_{a,N2}$) and ln $S_{N2/C}$ of spinel catalysts is uncovered in Fig. 3.3. This can be explained as follows. If $E_{a,N2} \approx E_{a,CO2}$ ($=E_a$) as in present results (Table 3.1), then $S_{N2/C} \approx (\Gamma A)_{N2}/(\Gamma A)_C$. Taking into account the compensation effect, $\ln \Gamma A = \gamma E_a + \text{const.}$, $S_{N2/C}$ is proportional to E_a. Thus, it can be concluded that spinels with larger E_a values for N_2 show higher selectivity to NO_x reduction into N_2.

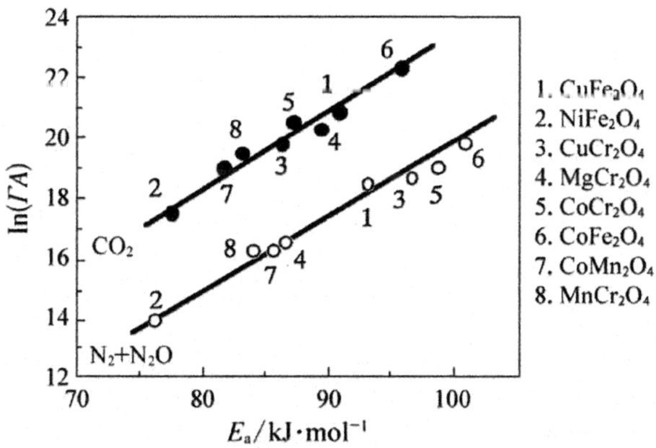

Fig. 3.3 Relation between the selectivity to $S_{N2/C}$ and $E_{a,N2}$

3.1.2 *Kinetic of Simultaneous NO_x–Soot Removal Over $CuFe_2O_4$*

In this section, the reaction orders with respect to reactant are quantified over $CuFe_2O_4$ for soot–O_2, soot–NO, and soot–O_2–NO reaction. Based on logarithms of the rate expressed by $r = kC_1^{n1}C_2^{n2}C_3^{n3}$, the equation is transformed into $\ln r = \ln k + n_1 \ln C_1 + n_2 \ln C_2 + n_3 \ln C_3$. A plot of $\ln r$ against $\ln C_1$ with C_2 and C_3 being held constant is supposed to be a straight line with slope n_1. Similarly, the order of n_2 or n_3 can be determined by plotting $\ln r$ against $\ln C_2$ or $\ln C_3$ with the fixed concentration of other reactants.

3.1.2.1 Soot–O_2 Reaction

The oxidation of soot particle without the assistance of catalysts requires high temperature (700 °C), yet it can be accomplished at relatively low temperatures (300–400 °C) over metal oxides catalysts [3]. In the present study, the soot–O_2 reaction is investigated over $CuFe_2O_4$ with the purge of gaseous O_2 (0.68, 2.41, 5.0%) diluted with He in the air. CO_2 is the solely detected product. Within the temperature range (300–400 °C), where the sufficient amount of charged soot remains, and kinetic analysis is possible, the ln–ln plot of CO_2 formation rate versus O_2 concentration is of linear relation with a slope of 0.46 (Fig. 3.5). This suggests the reaction is roundly a half order with respect to O_2 concentration. This is consistent with that of soot combustion catalyzed by Cu–V–K [4]. The half order can be interpreted if the following mechanism is assumed.

$$O_2 \underset{k_{-1}}{\overset{k_1}{\longleftrightarrow}} 2O_a \tag{3.3}$$

$$C + O_a \xrightarrow{k_2} [C * O] \tag{3.4}$$

$$[C * O] + O_a \xrightarrow{k_3} CO_2 \tag{3.5}$$

$$[C * O] + 0.5O_2 \xrightarrow{k_3} CO_2 \tag{3.6}$$

where O_a is the atomic oxygen adsorbed on the surface of the catalyst, [C * O] an intermediate possibly related to the attack by the O_a species on the activated carbon surface. The [C * O] intermediate then reacts with O_a (3.5) or gaseous O_2 (3.6) to produce CO_2. If the reaction (3.4) is the rate-determining step and reaction (3.3) is equilibrium with the equilibrium constant of

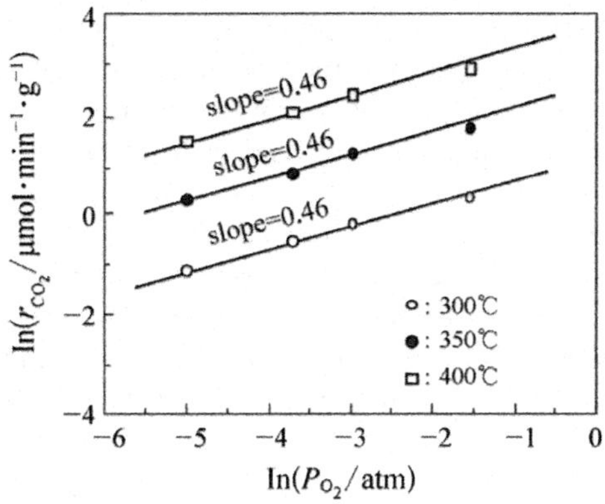

Fig. 3.4 Relation between ln r_{CO2} and ln P_{O2} in soot–O_2 reaction over $CuFe_2O_4$

$$K = [O_a]^2 / P_{O_2} \tag{3.7}$$

then the rate of CO_2 formation is given by (Fig. 3.4)

$$\frac{d[CO_2]}{dt} = k_2[O_a] = k_2 K^{0.5} P_{O_2}^{0.5} \tag{3.8}$$

3.1.2.2 Soot–NO Reaction

The reaction profiles are shown in Fig. 3.5 with various NO concentration feed. In two-component reaction, the reaction should be written as

$$C + 2NO \longrightarrow CO_2 + N_2 \tag{3.9}$$

and CO_2 and N_2 formed should be equal in molar amount. Actually, from TPO results, the two products are not always equal in molar ratio. The temperature

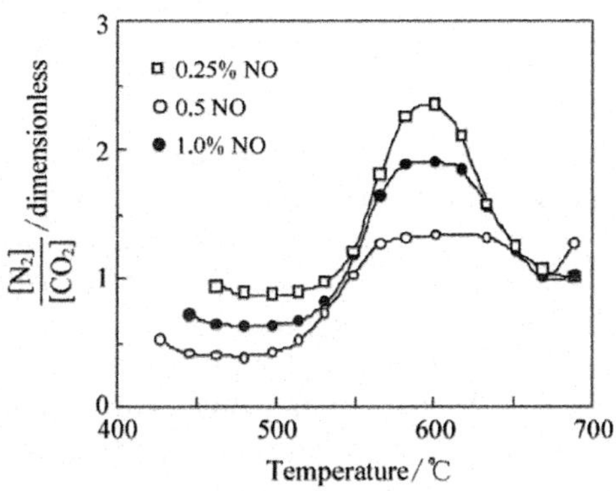

Fig. 3.5 Temperature dependence of the ratio of $[N_2]/[CO_2]$ for soot–NO reaction

dependence of the ratio $[N_2]/[CO_2]$ in various NO concentrations are presented in Fig. 3.5. The ratio indicates that an intermediate giving N_2 is accumulated on the catalyst or soot at lower temperatures, followed by the decomposition of accumulated intermediates and yield of N_2 at higher temperatures.

The reaction order concerning NO concentration has been examined below 462 °C utilizing the rate of CO_2 formation; as stated above, the rate of N_2 formation involves the effect of N-containing intermediate in this temperature range. Calculated from Fig. 3.7, the reaction rate is first order at low temperature, and this value tends to increase at higher temperatures. With the aim of explaining these findings, the following mechanism of the soot–NO reaction is proposed.

$$\mathrm{NO} \xrightarrow{k_1} \mathrm{NO_a} \tag{3.10}$$

$$\mathrm{NO_a} + \mathrm{C} \xrightarrow{k_2} [\mathrm{C,N,O}]^* \tag{3.11}$$

$$[\mathrm{C,N,O}]^* + \mathrm{NO_a} + 2\mathrm{C} \xrightarrow{k_3} \mathrm{CO_2} + 2[\mathrm{C,N}]^* \tag{3.12}$$

$$2[\mathrm{C,N}]^* \xrightarrow{k_4} 2\mathrm{C} + \mathrm{N_2} \tag{3.13}$$

NO_a is adsorbed NO on the catalyst surface, and $[]^*$ intermediates containing elements shown in the brackets. The alternative to reaction (3.12) and (3.13) is

$$[\mathrm{C,N,O}]^* + \mathrm{NO_a} \longrightarrow \mathrm{CO_2} + \mathrm{N_2} \tag{3.14}$$

It is evident that reaction (3.14) is not the present case since the molar ratio of CO_2 and N_2 is not identical. In the event that the adsorption of NO (3.10) or surface reaction between the intermediate and the adsorbed NO (3.13) is rate determining, the first-order (3.15) or the second-order kinetics with respect to NO concentration is acquired.

$$\frac{\mathrm{d}[\mathrm{CO_2}]}{\mathrm{d}t} = k_1 \mathrm{P_{NO}} \tag{3.15}$$

$$\frac{\mathrm{d}[\mathrm{CO_2}]}{\mathrm{d}t} = k_3 \mathrm{K_1^2 K_2 P_{NO}^2} \tag{3.16}$$

K_1 and K_2 are equilibrium constant of reactions (3.15) and (3.16), respectively. It can be deduced that soot–NO reaction is controlled by the NO adsorption step at lower temperatures. With the increase of heating temperature, the second-order kinetics (3.16) is gradually involved in addition to first-order kinetics, which gives the apparent reaction order between 1 and 2.

It should be highlighted that the above mechanism of the soot–NO reaction is tentative. That is, the participation of gaseous NO in steps of CO_2 formation and N_2 formation is not excluded. Thus, the overall reaction can be rewritten as

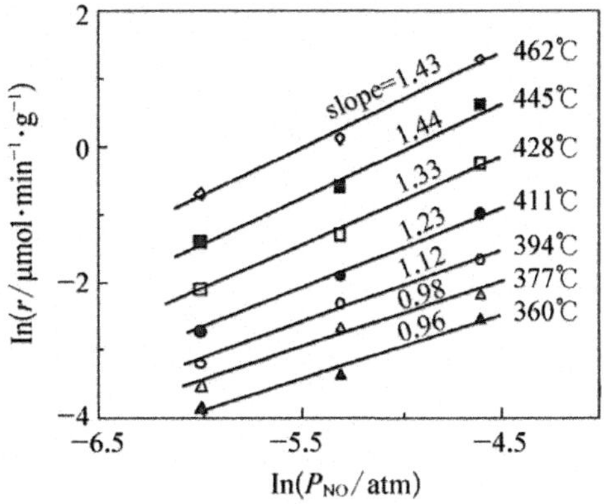

Fig. 3.6 Relation between ln r_{CO2} and ln P_{NO} in soot–NO reaction over $CuFe_2O_4$ [6]

$$[C, N, O]^* + NO + 2C \longrightarrow CO_2 + 2[C, N]^* \quad (3.17)$$

If this mechanism dominants the reaction, second-order kinetics with respect to NO concentration are also obtained. Similarly, also probable instance is the participation of gaseous NO in the N_2 formation step. Besides, the [C, N] intermediates have been reported in many kinds of literature for this process [5], which suggests that the assumption above is rational (Fig. 3.6).

3.1.2.3 Soot–O_2–NO Reaction

Effect of O_2 Concentration

In this reaction, the NO concentration is fixed; the variable O_2 concentration (1.5, 3.0, 5.0%) is adopted, and the N_2 and CO_2 formed throughout the TPO process are detected. As can be seen from the ln–ln plots of O_2 concentration versus the rate of N_2 formation (Fig. 3.7a), the rate of N_2 formation is about the 0.4th order with respect to O_2 concentration in the temperature range from 290 to 360 °C. As listed in Table 3.2, the effect of O_2 on the N_2O formation is similar to that of N_2 formation rate (0.4th order), and the CO_2 formation rate is about the 0.6th order with respect to O_2 concentration.

Effect of NO Concentration

The reaction order with respect to NO concentration is measured in NO (0.5, 0.3, 0.225, and 15%)-O_2 (5%)-He balance. The ln–ln plots of N_2 formation rate versus NO concentration are presented in Fig. 3.7b and are linear with a slope of ca. 1.0. The CO_2 and N_2O formation rates are 0.6th and the first-order dependence on NO concentration, respectively.

The Arrhenius plots for formation rates of CO_2 and N_2 in different O_2 and NO concentrations are displayed in Figs. 3.8 and 3.9, respectively. It reveals the secondary effect of NO concentration on apparent activation energy to that of O_2 concentration.

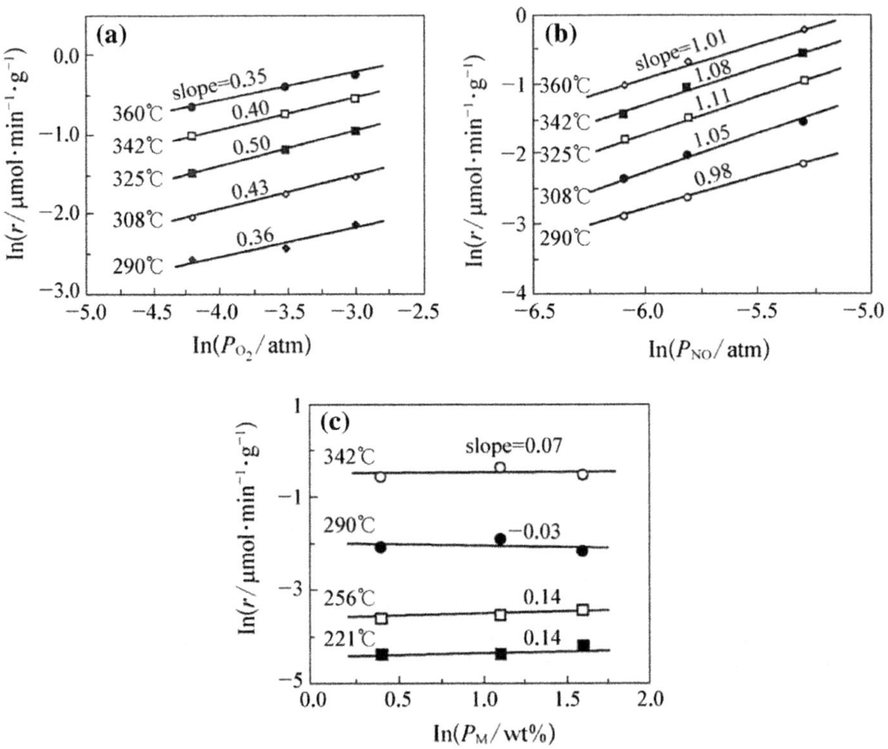

Fig. 3.7 Influence of concentrations of reactants on the rate of N_2 formation in the simultaneous NO–soot removal over $CuFe_2O_4$, **a** influence of O_2 concentration; **b** influence of NO concentration; **c** influence of the charged amount of soot particle

Table 3.2 Reaction orders with respect to the concentration of O_2, NO, and amount of charged soot in the simultaneous NO_x–soot removal over the $CuFe_2O_4$ catalyst

Reaction rate	Temp. (°C)	Reaction order		
		O_2	NO	Soot
r_{CO2}	290	0.54	0.58	−0.03
	308	0.58	0.60	0.07
	325	0.61	0.60	−0.02
	342	0.59	0.55	−0.10
r_{N2}	290	0.36	0.98	0.13
	308	0.43	1.05	−0.10
	325	0.50	1.11	0.15
	342	0.40	1.08	0.07
r_{N2O}	359		1.11	
	394	0.48	0.92	
	411	0.40	0.94	

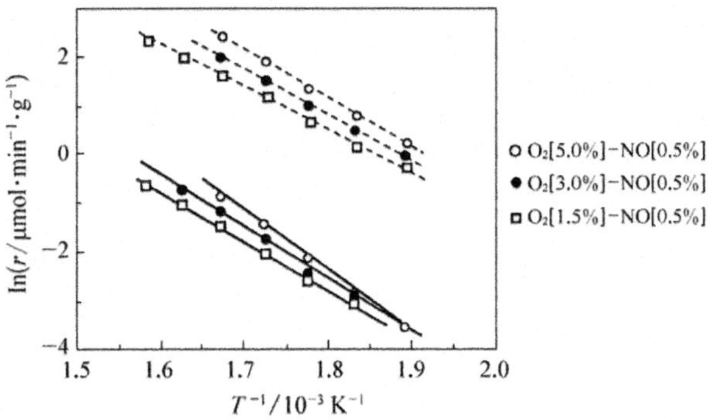

Fig. 3.8 Arrhenius plots for formation rates of N_2 (solid line) and CO_2 (broken line) in the NO–O_2 atmosphere with various O_2 concentration over $CuFe_2O_4$

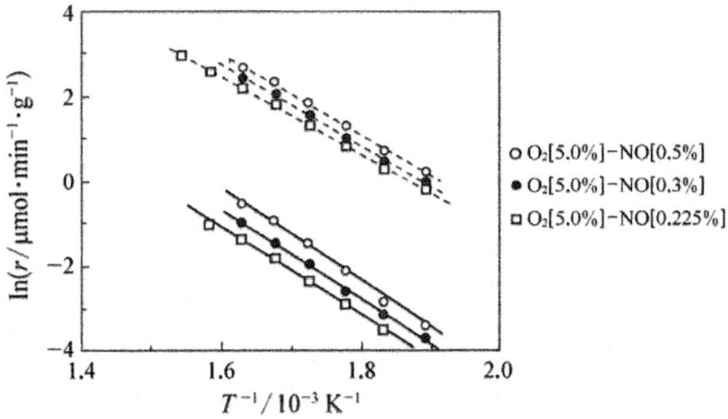

Fig. 3.9 Arrhenius plots for formation rates of N_2 (solid line) and CO_2 (broken line) in the NO–O_2 atmosphere with various NO concentration over $CuFe_2O_4$

Effect of Charged Amount of Soot

The rate of N_2 formation is not affected by the soot/catalyst ratio as shown in Fig. 3.7c; the order of reaction with respect to soot concentration is 0. This is quite natural since the kinetic analysis is performed in the range where the soot/catalyst contact area is regarded as approximately constant. As the conversion is at a low level, this presumed variable (contact area) remains unchanged.

Expressions of Reaction Rates and Selectivity to N_2 Formation

According to the results obtained above, the rates of CO_2, N_2, and N_2O formation in the simultaneous NO_x–soot removal reaction can be expressed as follows; all the rates are zero order with respect to the amount of soot.

$$r_{CO_2} = k_1[O_2]^{0.6}[NO]^{0.6} \quad (3.18)$$

$$r_{N_2} = k_2[O_2]^{0.4}[NO]^{1.0} \quad (3.19)$$

$$r_{N_2O} = k_2[O_2]^{0.4}[NO]^{1.0} \quad (3.20)$$

The relation between the concentration of reactants and the reaction rate implies that the mechanism of the simultaneous NO_x–soot removal is sophisticated. From

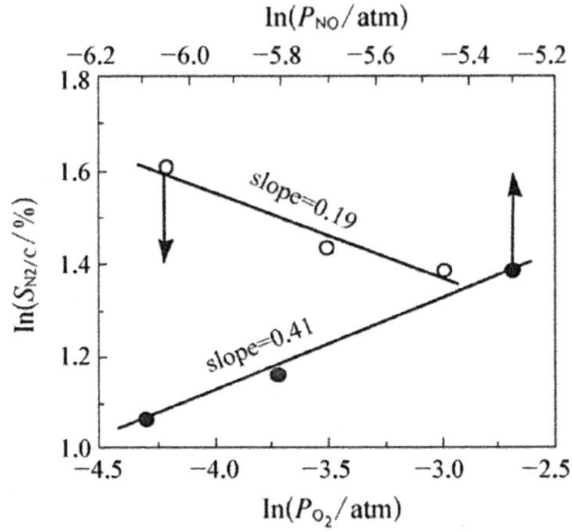

Fig. 3.10 ln–ln plots of selectivity to the nitrogen formation ($S_{N2/C} = r_{N2}/r_{CO2}$) versus concentration of O_2 (open circle) and NO (closed circle) in the simultaneous NO_x–soot removal

the rate expression (3.18) and (3.19), the selectivity to N_2 formation can be expressed by (Fig. 3.10)

$$\alpha = r_{N_2}/r_{CO_2} = k'[O_2]^{-0.2}[NO]^{0.4} \tag{3.21}$$

Effect of Gaseous Composition

The reaction rates in various gas compositions are summarized in Fig. 3.11 as an Arrhenius plot. This indicates that the activity for soot combustion (CO_2) formation is strongly dependent on reactant gas and increases as the sequence: NO < O_2 < NO + O_2 < NO_2 + O_2 < NO_2. The activity for N_2 formation also enhanced as NO < NO + O_2 < NO_2 + O_2 < NO_2. Thus, in the presence of NO_2, the reaction $2NO_2 + C = 2NO + CO_2$ proceeds at a lower temperature prior to the initiation of N_2 yielding; the Arrhenius correlation line of N_2 generation in NO_2 atmosphere shifts to higher temperatures and come close to that of N_2 formation in NO + O_2.

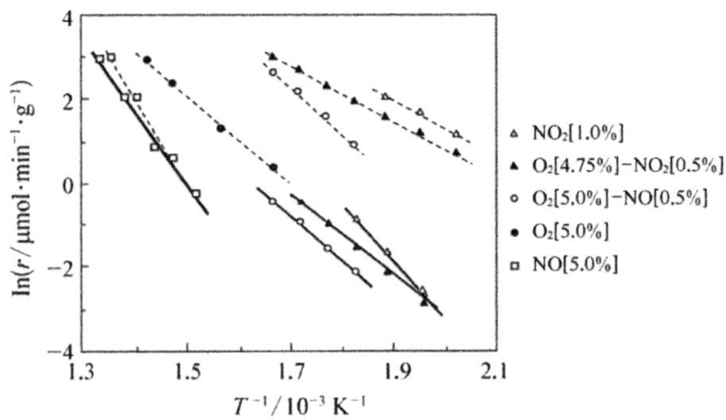

Fig. 3.11 Comparison of rates of CO_2 (broken line) and N_2 (solid line) formation over $CuFe_2O_4$ in O_2, NO_2, NO–O_2, and NO_2–O_2 reaction gasses

Table 3.3 Comparison of the ignition temperature of soot and apparent activation energy in various reactant gasses

Reaction gases	$T_{ig}^{①}$ (°C)	$E_{a,N2}^{②}$ (kJ mol^{-1})	$E_{a,CO2}^{③}$ (kJ mol^{-1})
NO_2 [1.0%]	195	123	57.4
NO_2 [0.5%] + O_2 [4.75%]	250	83.3	52.6
NO [0.5%] + O_2 [5%]	288	91	90.6
O_2 [5%]	373	n/a	88.9
NO [5%]	418	147.4	170.6

①Ignition temperature of the soot
②Apparent activation energy of N_2
③Apparent activation energy of CO_2

Table 3.3 lists the values of T_{ig}, $E_{a,N2}$, and $E_{a,CO2}$ in various reactant gas which significantly influence not only the T_{ig} and E_a. T_{ig} depends on the reaction gas as NO < NO + O_2 < NO_2 + O_2 < NO_2, which is in accord with the statement above. In terms of E_a, on the other hand, E_a for CO_2 formation in NO_2 and NO_2–O_2 is lower than that for N_2 formation; the reverse is true with NO gas, and both E_a are close to each other in NO–O_2 gas. The activation energy of CO_2 formation in NO_2 gas (57.4 kJ/mol) is close to that in NO_2–O_2 (52.6 kJ/mol), and they are far much lower than that in O_2 (88.9 kJ/mol). This indicates that soot is oxidized by nitrogen dioxide instead of oxygen in the NO_2 + O_2 atmosphere.

Reaction Mechanism

The reaction mechanism can be inferred according to the clues for kinetics study, and it is presented in Fig. 3.12. In this section, the pathways for the reaction are debated especially taking account of the participation of NO_2 and the effect of coexisting O_2.

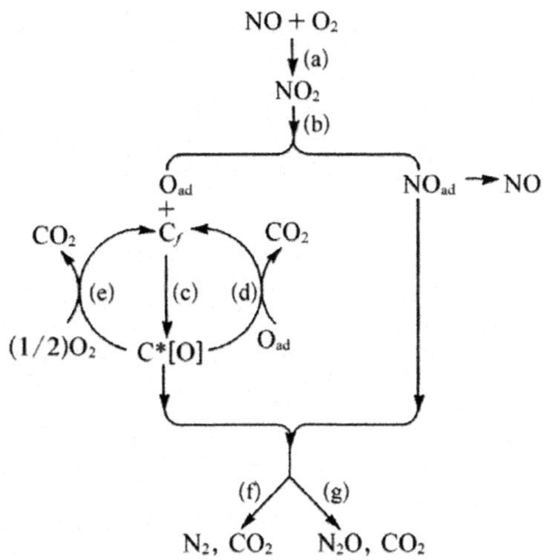

Fig. 3.12 Proposed reaction scheme of simultaneous NO_x–soot removal with the introduction of O_2 [7]

Judging by the T_{ig} value in reactant gasses ($NO_2 < NO_2 + O_2 < NO + O_2 < O_2 < NO$), the reaction in the NO + O_2 gas stream is boosted by NO_2 which is formed by gas-phase oxidation of NO via O_2 (a). NO_2 molecules are then dissociatively adsorbed on the catalyst surface (b). As can be recognized from the fact that NO and CO_2 are the products in the NO_2–soot reaction at lower temperatures, NO_a species is less reactive and readily desorbs as molecular NO, while O_a species is reactive with soot and evolve CO_2. It is assumed that the adherence of O_a to soot to form carbon–oxygen intermediate (C * O) is of higher priority (c). The activated C * O species then reacts with O_a and produce CO_2. It is noteworthy that the removal of one surface carbon atom generates some active carbon species which have bound to the removed carbon atom; hence, the reaction (d) is likely to be plausible for this step. In addition to the oxidation of NO to NO_2, coexisting O_2 plays another role in increasing the surface of active carbon sites. One possibility is that O_2 molecules are dissociatively adsorbed to give O_a species which participates in reaction (c) and (d). In the temperature range (below 340 °C) for kinetic study, the soot–O_2 reaction is unlikely to dominate the NO_x–soot reaction judging by the synchronism of CO_2 profiles and N_2/N_2O curves. Instead, gaseous O_2 directly attacks activated carbon to form C * O and CO_2 (e). N_2 formation in NO–O_2–soot is quite low, reaction (e) is deemed to be a fast process followed by the contribution to the evolution of active surface carbon sites. As the concentration of active carbon sites dense, there's more opportunity to reacts with NO_a to generate N_2 and N_2O eventually (f) and (g). It is underlined that the reaction (f) and (g) are simplified; N-containing intermediate is to be involved in the overall reaction pathway [7].

3.1.3 Kinetic of Simultaneous NO_x–Soot Removal Over K-Substituted $CuFe_2O_4$

3.1.3.1 Kinetic of Simultaneous NO_x–Soot Removal Over $Cu_{0.95}K_{0.05}Fe_2O_4$

It has been presented in the last chapter that $Cu_{0.95}K_{0.05}Fe_2O_4$ exhibits a promotion effect on both catalytic activity and selectivity. It is expected that the reaction order over K-substituted catalysts should be different from that over unsubstituted. The effect of O_2 and NO concentration on reaction rates of simultaneous NO–soot removal over $Cu_{0.95}K_{0.05}Fe_2O_4$ was investigated. The summarized results are displayed in Table 3.4, and the expression of reaction rates can be simplified as:

$$r_{CO_2} = k_1[O_2]^{0.8}[NO]^{0.8} \quad (3.22)$$

$$r_{N_2} = k_2[O_2]^{0.7}[NO]^{1.4} \quad (3.23)$$

Table 3.4 Reaction order with respect to the concentration of O_2 and NO in the simultaneous NO_x–soot removal over $Cu_{0.95}K_{0.05}Fe_2O_4$ catalyst

Reaction rate	Temp. (°C)	Reaction order	
		O_2	NO
r_{CO2}	273	0.73	0.87
	290	0.80	0.94
	307	0.83	0.88
	324	0.81	0.71
r_{N2}	273	0.63	1.52
	290	0.74	1.46
	307	0.81	1.33
	324	0.73	1.13
r_{N2O}	290		1.02
	307	0.98	1.23
	324	0.95	1.05

$$r_{N_2O} = k_3[O_2]^{1.0}[NO]^{1.1} \quad (3.24)$$

then the selectivity to N_2 formation can be expressed as:

$$S_{N_2/C} = r_{N_2}/r_{CO_2} = k'[O_2]^{-0.1}[NO]^{0.6} \quad (3.25)$$

The comparison between Eqs. (3.21) and (3.25) uncovers that the selectivity to N_2 formation over $Cu_{0.95}K_{0.05}Fe_2O_4$ is much more positively dependent on NO concentration and much less negatively depend on O_2 concentration than that over pure $CuFe_2O_4$.

3.1.3.2 Effect of the Substitution Amount of Potassium and Reaction Gas Composition

In Table 3.5, the comparison between K-substituted and K-free catalysts is made on T_{ig} and selectivity parameters in various $[O_2]$/NO ratios. In all the atmospheres, the K incorporation leads to not only the significant enhancement of the activity and the selectivity to N_2 formation ($S_{N2/C}$) but also the slight decrease in the selectivity to nitrous oxide (S_{N2O}). The effects of O_2 and NO concentrations on the apparent activation energy are shown in Figs. 3.13 and 3.14, respectively. The K-substitution increased the apparent activation energies of CO_2, N_2, and N_2O significantly.

Figure 3.15 shows the effect of K-substitution level on T_{ig} in $Cu_{1-x}K_xFe_2O_4$. In each substitution level (x), T_{ig} depends on the gas composition as $O_2 + NO < O_2 < NO$, which can be accounted for the sequence of gas oxidability. The T_{ig} in NO + O_2 gas achieves minimum at $x = 0.05$, as stated in Chap. 2. In O_2 or NO gas, on the other hand, T_{ig} shows a decreasing tendency with the increase of x, through its dependence of x somehow differs between the two; monotonic up to 0.15 and 0.05 in O_2 and NO atmospheres, respectively, followed by the leveling off. The minimum T_{ig} in NO + O_2, as well as the x-dependence of T_{ig} in NO + O_2 gas

Table 3.5 Effect of K-substitution for Cu in $Cu_{1-x}K_xFe_2O_4$ on catalytic performance in various gaseous compositions

Gas	$T_{ig}^{①}$ (°C)		$S_{N2/C}^{②}$ (%)		$S_{N2O}^{③}$ (%)	
(O_2/NO)	$x = 0$	$x = 0.05$	$x = 0$	$x = 0.05$	$x = 0$	$x = 0.05$
5.0/0.5	285	266	4.7	5.9	20.8	18.6
5.0/0.3	305	270	3.2	4.4	22.8	21.1
5.0/0.225	320	275	2.5	3.2	22.4	21.4
3.0/0.5	304	270	4.9	6.2	23.1	16.4
1.5/0.5	323	275	6.9	7.1	15.3	11.0

①Ignition temperature of soot
②Selectivity to N_2 formation
③Selectivity to the N_2O formation

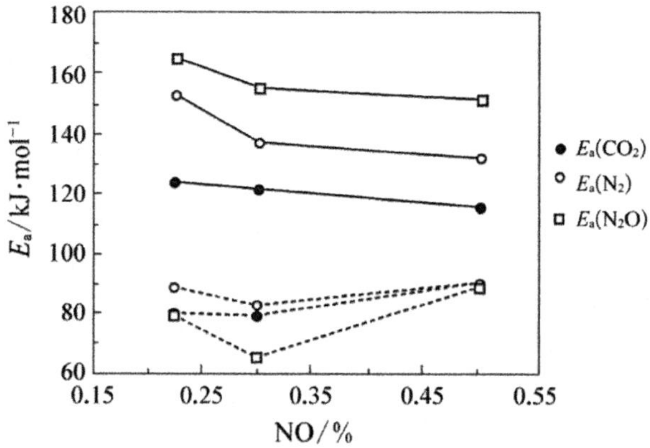

Fig. 3.13 Apparent activation energy (E_a) of CO_2, N_2, and N_2O formation as a function of NO concentration in soot–O_2 [5%]-NO reaction over $CuFe_2O_4$ (broken line) and $Cu_{0.95}K_{0.05}Fe_2O_4$ (solid line)

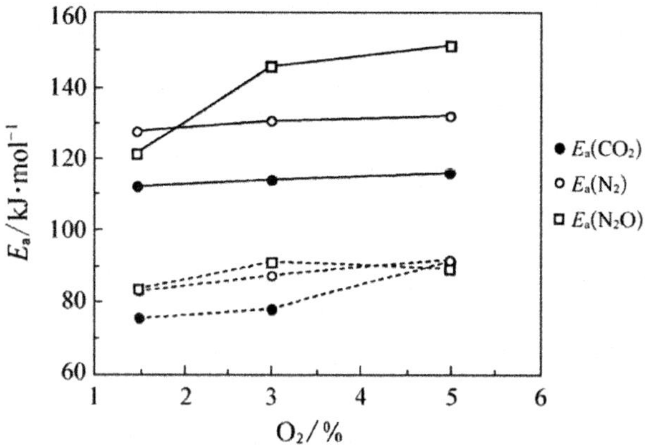

Fig. 3.14 Apparent activation energy (E_a) of CO_2, N_2, and N_2O formation as a function of NO concentration in soot–O_2–NO[0.5%] reaction over $CuFe_2O_4$(broken line) and $Cu_{0.95}K_{0.05}Fe_2O_4$ (solid line)

which is different from that in O_2 or NO gas, may suggest that, in NO + O_2, NO_2 participates in the reaction.

In terms of compensation effect, the linear relations between E_a and ln ΓA over K-substituted samples are acquired (Fig. 3.16b), which implies reactant gas plays a crucial role in the mechanisms and kinetics.

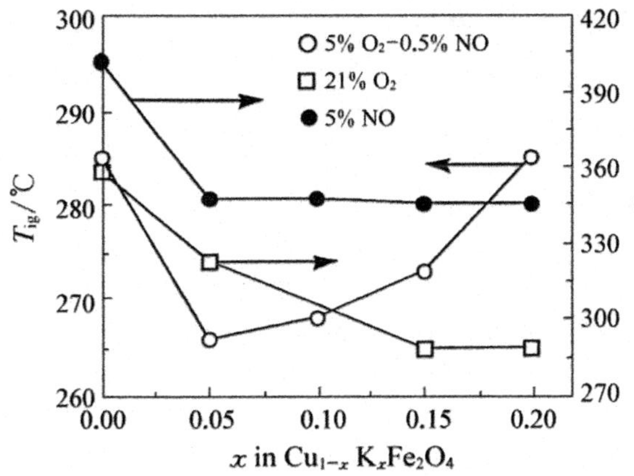

Fig. 3.15 Effect of potassium substitution for Cu in $Cu_{1-x}K_xFe_2O_4$ on the T_{ig} in various atmospheres

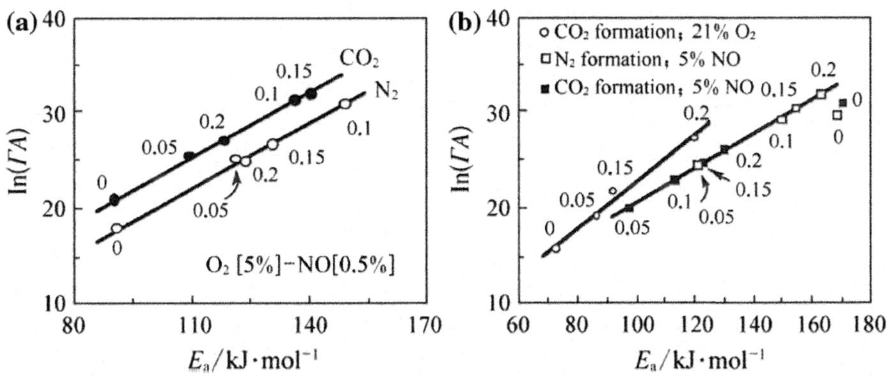

Fig. 3.16 Compensation effect in CO_2, N_2 formation over $Cu_{1-x}K_xFe_2O_4$ in the gas mixture of **a** O_2 [5%]-NO-[0.5%], **b** O_2 [21%]-NO-[5%]. Numerical values of x in $Cu_{1-x}K_xFe_2O_4$ are denoted in the figure

Based on the above results, the reaction pathways for K-substituted $CuFe_2O_4$ samples are further interpreted. Since potassium substitution favors both T_{ig} and N_2 formation, it is believed that K will promote reaction (b) and (f).

3.2 Numerical Model of NO_x–Soot Removal

The overall NO_x–soot can be expressed as follows:

$$C + NO + O_2 \longrightarrow CO_2 + N_2 + N_2O \tag{3.26}$$

3.2.1 Mathematical Model

The chemical kinetic equation can be written as

$$r_{i,CO_2} = A_i \cdot \exp\left(-\frac{E_i}{RT}\right) \cdot S_i \cdot P^{m}_{O_2} P^{n}_{NO} \tag{3.27}$$

r_{i,CO_2} is the reaction rate; *I* is the pre-exponential factor; E_i represents the activation energy; and *n* is the reaction order for O_2 and NO, respectively; S_i is the surface area of soot. Its relation to the initial surface area of soot (S_{i0}) can be expressed as follows:

$$S_i = S_{i0}(1 - X_i)^k \tag{3.28}$$

X_i represents the soot conversion while *k* is the reaction order for soot. *k* value depends on the morphology of soot particles and the consumption amount during the oxidation process. If all the particles can be adjacent to the catalyst, then $k = 1$; If the soot particle is in a sphere shape, then $k = 2/3$.

In terms of spheres, the ratio of surface area to volume is fixed at the value of 3/ *R*.

$$S_{i0} = \frac{3}{R_{i0}} \cdot V_{i0} = \frac{3 \cdot C_{i0}}{R_{i0} \cdot d_c} \tag{3.29}$$

V_{i0} and R_{i0} are the volume and radius of soot particle, respectively, before the reaction; C_{i0} is the initial soot concentration (soot/catalyst). The following expression is acquired if we combine Eqs. (3.28) and (3.29) into Eq. (3.27).

$$r_{i,CO_2} = A_{is} \cdot \exp\left(-\frac{E_i}{RT}\right) \cdot C_{i0} \cdot (1 - X_i)^k \cdot P^{m}_{O_2} \cdot P^{n}_{NO} \tag{3.30}$$

$$A_{is} = \frac{A_i \cdot 3}{R_{i0} \cdot d_c} \tag{3.31}$$

Then, the concentration of CO_2 can be written as a function as $r_{i,CO2}$ as follows:

$$C_{ti} = \frac{r_{i,CO_2} \cdot w}{F_0} \tag{3.32}$$

F_0 is the flow rate and w is the mass of catalyst. All the other parameters are acquired in this chapter; hence, C_{ti} is a function of T. It is possible to model the reaction profiles.

3.2.2 *Simulation*

According to two groups of comparison depicted in Figs. 3.17, 3.18, 3.19, and 3.20, the simulated results perfectly fit the experimental patterns in both $NO + O_2$ and O_2 conditions, which show this numerical model is based on logical assumption.

The numeric model provides access to reproduce the reaction and might be another methodology to screen the catalyst.

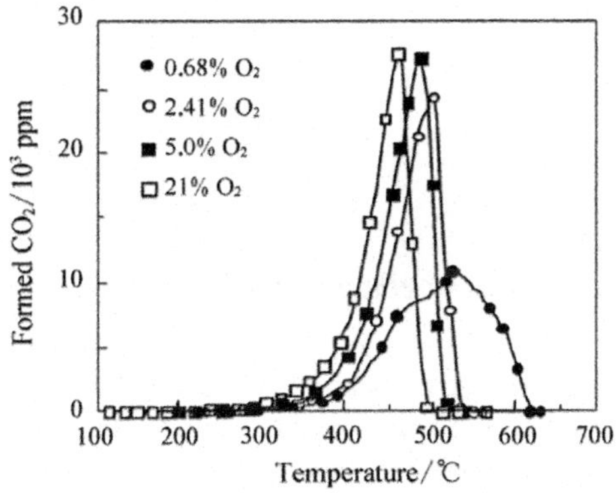

Fig. 3.17 Reaction profiles for soot–O_2 reaction over $CuFe_2O_4$ with various O_2 concentrations

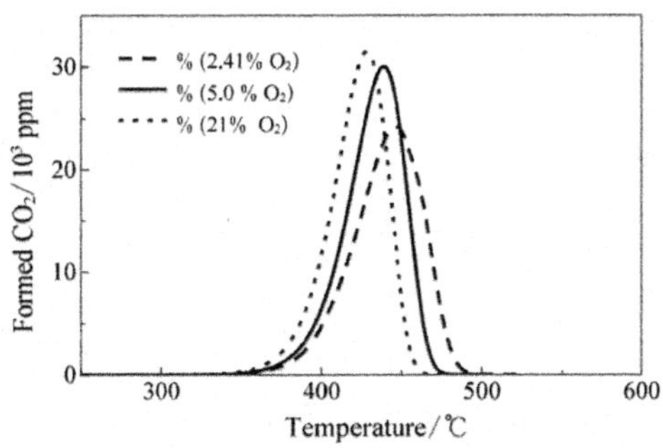

Fig. 3.18 Simulated results of soot–O_2 reaction over $CuFe_2O_4$ in various O_2 concentrations

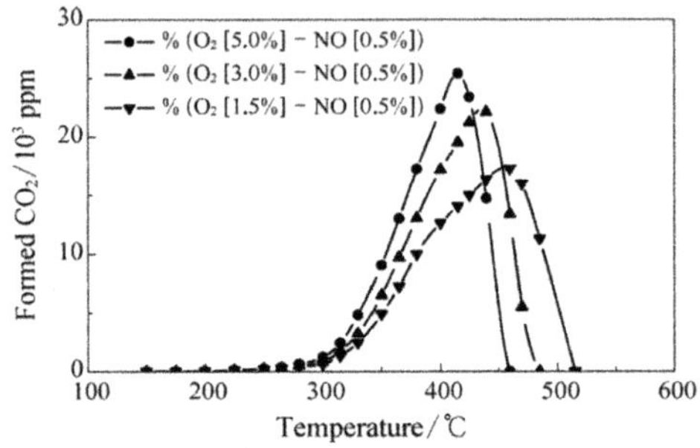

Fig. 3.19 Reaction profiles for soot–NO–O_2 reaction over $CuFe_2O_4$ with various O_2 concentrations

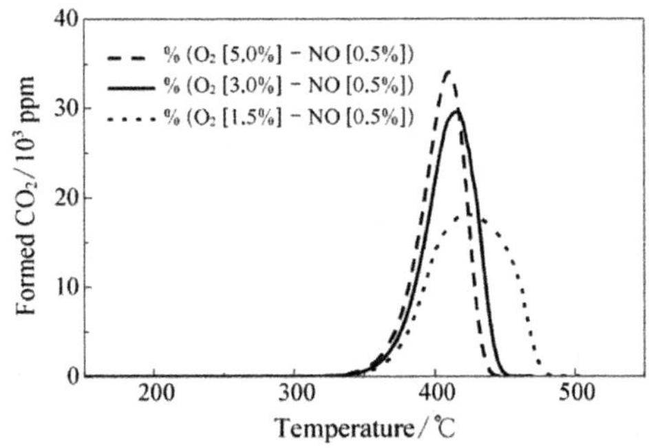

Fig. 3.20 Simulated results of soot–O_2 reaction over $CuFe_2O_4$ in various O_2 concentrations

3.3 Theoretical Study of PM–NO_x Catalytic Reaction

Investigating the nature of the interaction of NO_x with the surface of soot particles together with the catalysts by theoretical means could be beneficial for a better understanding of some aspects of the simultaneous removal processes and can be seen as complementary to the valuable data collected experimentally. However, judging by the fact that the simultaneous removal reaction takes place at the solid (soot)–solid (catalyst)–gas triple-phase conjunction, it will be confusing to model the soot/catalysts surface. Thus in this part, we will briefly introduce current research focused on the non-catalytic NO_x–soot reaction and theoretical research involved NO adsorption on the catalyst surface. As a preliminary step toward a better understanding of the catalytic NO_x and soot reaction, we hope this brief review could shed light on the heterogeneous simultaneously removal catalysis reaction of NO_x on the soot surface.

3.3.1 Insight into the Interaction Between NO and Soot

In a series of recent papers, theoretical approaches are based on either classical simulation method or first-principle calculation to characterize the interaction between carbonaceous surfaces and NO_x without catalysts.

The most common choice is to model the oxidized surface by introducing oxygen-containing functionalities (such as carboxylic acids, ketones, ethers, and alcohols) in molecular polycyclic aromatic hydrocarbon (PAH-) type systems, but periodic graphene layers (with or without defects) are also employed.

Markus Kraft carried out a theoretical study on the interaction between soot and NO molecules in the absence of oxygen in order to further develop the mechanistic understanding behind the formation of chemical species such as CO, N_2, and N_2O on soot [8]. An example PAH, showing the principal surface site types is showed in Fig. 3.21. These reactions take place on different types of reactive sites (e.g., armchair sites, bay sites, and embedded and surface N radicals). Seven different reaction pathways on PAH molecule have been presented. Two typical reaction pathways were shown in Figs. 3.22 and 3.23.

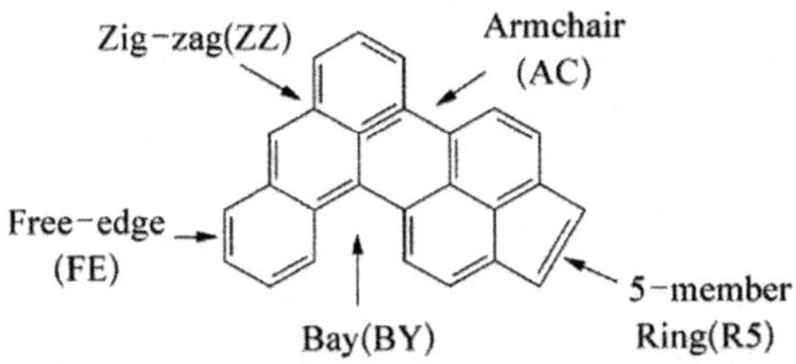

Fig. 3.21 An example PAH showing the principal surface site types [8]

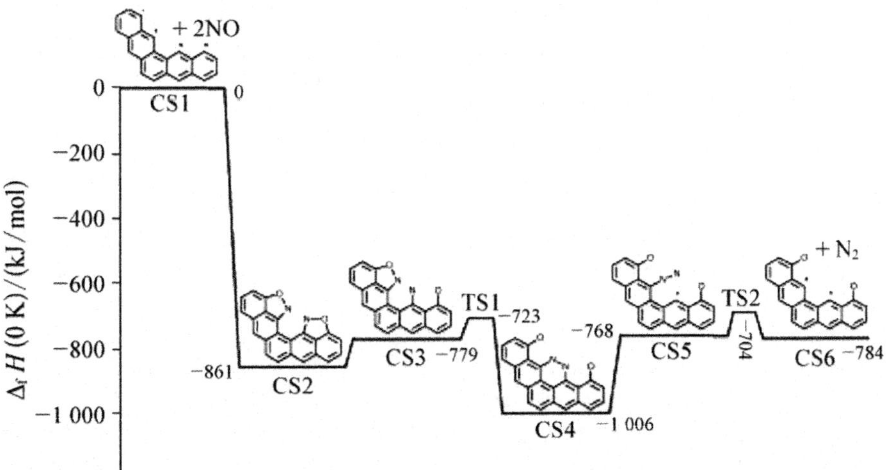

Fig. 3.22 Pathway 1: Potential energy diagram showing the formation of an N_2 molecule on a PAH, when two NO molecules are chemisorbed on two ZZ sites, and the two N atoms interact on an armchair site. For PAH reactions to take place through this pathway, an armchair site in between two ZZ sites are required (fulfilled by CS1) [8]

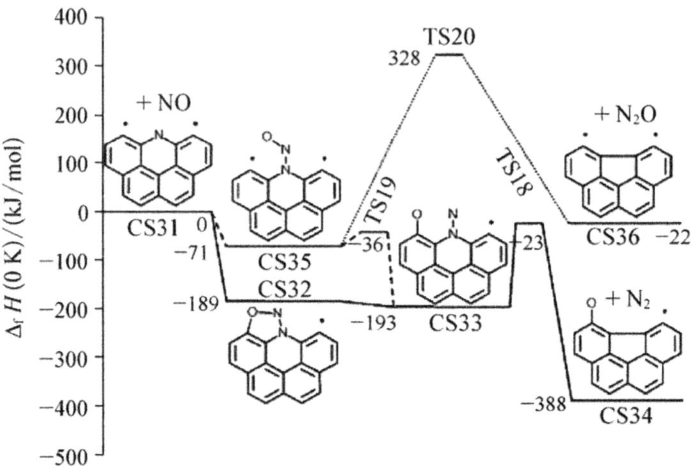

Fig. 3.23 Pathway 7: potential energy diagram showing the reactivity of embedded N atoms involved in the formation of 6-member rings on PAHs toward NO molecules [8]

The energetics and kinetics of the PAH reactions were evaluated using density functional theory and transition state theory, respectively. The chemisorption of NO on PAHs was found to be exothermic irrespective of the relative orientations considered in this work. The formation of N_2 on PAHs involved low energy barriers indicating that it can form at low temperatures as observed experimentally and a new rate for the removal of CO from PAHs was proposed.

A PAH growth model called the kinetic Monte Carlo—aromatic site (KMC—ARS) model has been used to simulate the formation of gas-phase species on PAHs by using the reactions proposed in this work appended with the reactions listed in. The model predicts the formation of CO in a substantial amount at temperatures >873 K using the rate proposed in this work, which is observed experimentally in soot–NO environments. The observed experimental trends in isothermal conditions that the amount of N atoms present on the soot surface increases and the number of O atoms decreases with increasing temperature have been seen in the simulations. Furthermore, it was observed in that 40–63% of N atoms remained on the soot surface after the soot–NO reactions, which agree reasonably well with the model prediction of 50–60%. However, the amount of CO and N_2 formed in the reactions was slightly underpredicted. There are many possible reasons behind this such as the presence of N and O atoms as impurities in soot, the presence of highly reactive sites such as ZZs leading to the faster conversion of NO to CO and N_2 or an incomplete reaction mechanism. Therefore, further development of the mechanism is required. The NO–soot reaction was found to depend strongly on the soot site type and temperature. For a set of temperatures, computed PAH structures were analyzed to determine the functional groups responsible for the decrease in the reactivity of soot with NO with increasing reaction time. In isothermal conditions, it was found that as the temperature is increased, the number of oxygen atoms remaining on the soot surface decreases, while the number of nitrogen atoms increases for given reaction time.

Picaudre investigates the adsorption of NO on perfect carbonaceous clusters, mainly by using DFT approaches [9]. They have shown that the interaction of NO with a perfect carbonaceous surface corresponds to a physisorption process that requires using a functional taking into account the dispersion effects. In this way, ωB97X-D represent an accurate functional for studying physisorption on weakly interacting systems such as those considered here. In the corresponding equilibrium geometry (Fig. 3.24), the NO radical is adsorbed nearly parallel to the carbonaceous surface, the N atom being slightly closer to the surface than the O atom (the tilt angle is equal to about 15° with respect to the carbon plane). In this position, the N and O atoms are located nearly above a C atom ("top" site) and above the center of a carbon hexagon ("hollow" site), respectively. The molecule–surface distance (N–C) is around 3.1 Å. The binding energy values obtained here with the ωB97X-D functional are consistent with the experimental data obtained at a relatively low temperature and submonolayer coverage, irrespective of the cluster size considered in the calculations. As a conclusion of this study on NO adsorption on the face of a carbonaceous cluster, this supports the hypothesis of a physisorption process for NO on C_nH_m clusters, mainly governed by the dispersion interactions. The

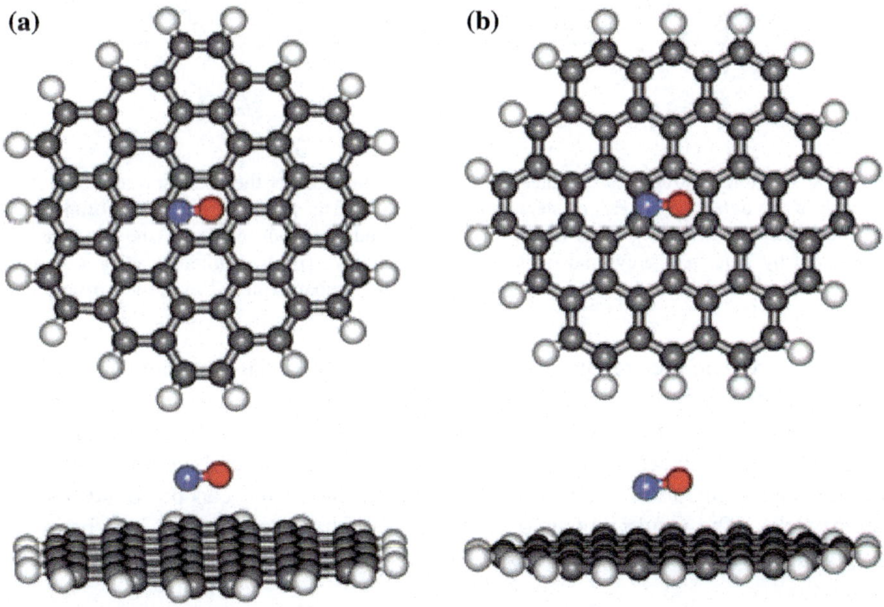

Fig. 3.24 Optimized structures calculated for the adsorption of NO on the face of a $C_{54}H_{18}$ cluster using the 6-31G basis set and the **a** ωB97X-D or **b** LSDA functional. Both top and side views of the system are shown. N, O, C, and H atoms are represented by blue, red, gray, and white balls, respectively [9]

Table 3.6 Binding energies (in kcal/mol) calculated for NO adsorbed at the defective edge of a $C_{54}H_{17}$ cluxxster [9]

	$C_{54}H_{17}$–NO singlet			$C_{54}H_{17}$–ON triplet		
	DFT		ab initio	DFT		ab initio
	ωB97X-D	B3LYP	HF	ωB97X-D	B3LYP	HF
Restricted and RO methods	−61.9	−59.3	−31.8	−30.0	−28.4	−34.9
Unrestricted	−60.3	−58.8	−24.2	−33.3	−27.8	−27.1
($\langle S^2 \rangle$ value for $C_{54}H_{17}$, NO, adsorbed system)	(0.78)	(0.76)	(1.33)	(0.78)	(0.762)	(1.33)
	(0.752)	(0.752)	(0.93)	(0.752)	(0.752)	(0.93)
				(2.00)	(2.009)	(2.07)

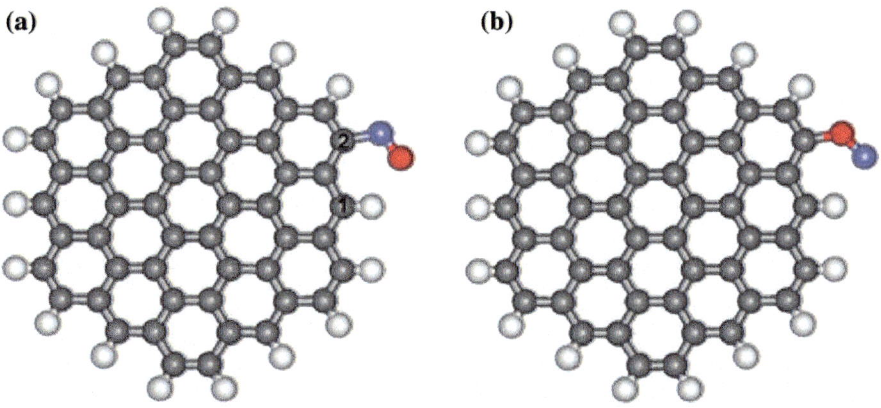

Fig. 3.25 Optimized structures calculated with the DFT method for the chemisorption of NO on the edge of the defective $C_{54}H_{17}$ clusters using the 6-31G basis set and the ωB97X-D functional [9]. Binding through either a C–N (**a**) or a C–O (**b**) bond is shown. N, O, C, and H atoms are represented by blue, red, gray, and white balls, respectively. The two different adsorption sites considered in the calculations (see text) are indicated by numbers 1 and 2 in the left panel

approach used here thus represents a significant improvement compared to previous studies on similar systems.

In a second step, they have used various theoretical approaches to characterize the interaction between NO and the defective edges of carbonaceous clusters because such a situation could also model the surfaces that can be found in soot nanoparticles. The results obtained in the $C_{54}H_{17}$ cluster with the 6-31G basis set are given in Table 3.6. The corresponding adsorption geometries are illustrated in Fig. 3.25.

In contrast with the adsorption on the face, the interaction of NO with the edge of a defective cluster leads to chemical bonding, with a net preference for a C–N rather than a C–O binding that can be explained on the basis of simple theoretical chemistry considerations.

The conclusion of this work is thus that soot may be an efficient sink for NO in the troposphere only if it contains a sufficiently high number of defective sites, mainly unsaturated carbon atoms. This work also clearly evidenced that quantum calculations (both at the DFT and HF levels of theory) have to be carefully conducted when considering the interaction between atmospheric species and carbonaceous surfaces, although these systems might appear quite simple at first glance. Thus, any credible conclusion based on quantum calculations for the mechanism of the simultaneous removal reaction for the NO_x and soot together with the catalyst should be drawn once systematically and carefully addressed the problems encountered with system size, functional type, size of the basis set, spin contamination, and spin multiplicity.

3.3.2 DFT Calculation for the Adsorption of NO Over $CuFe_2O_4$

The adsorption property is closely related with the nature of the catalysis, and knowledge of adsorbate adsorption may serve as a basis for the development of a comprehensive mechanism for the simultaneously catalytic removal process of nitrogen oxides and diesel particulates. Thus, the electronic properties of the spinel-type $CuFe_2O_4$ material and the adsorption behavior of NO molecule on

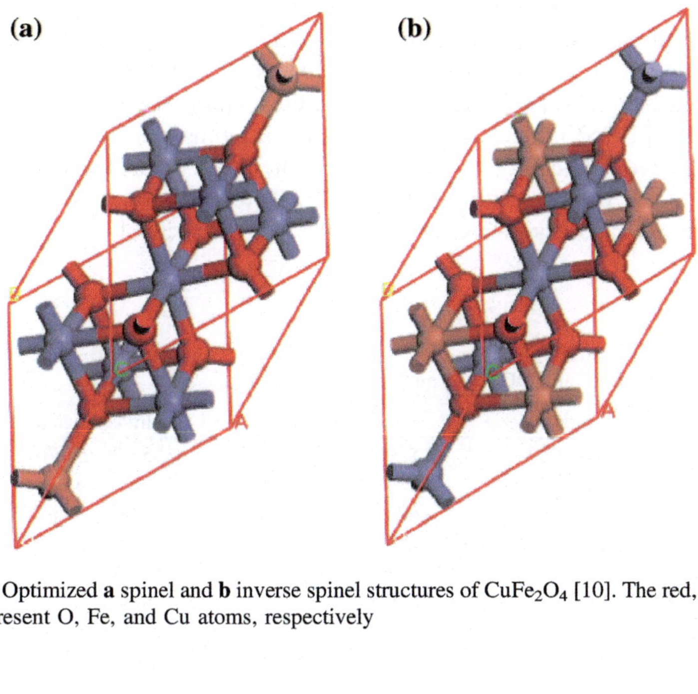

Fig. 3.26 Optimized **a** spinel and **b** inverse spinel structures of $CuFe_2O_4$ [10]. The red, green, and brown represent O, Fe, and Cu atoms, respectively

$CuFe_2O_4$ (100) surface were studied by using density functional theory method with on-site correction for Coulomb interaction (DFT + U) to provide an initial understanding to the catalysis of NO molecule over $CuFe_2O_4$ surface [10].

The optimized normal and inverse spinel structures of bulk $CuFe_2O_4$ are shown in Fig. 3.26. The standard $CuFe_2O_4$ spinel structure has a face-center-cubic (fcc) lattice with space group $Fd\bar{3}m$, where the tetrahedral hollow sites (A site) in an fcc close-packed oxygen sublattice are occupied by Cu atoms, and the octahedral hollow sites (B site) are occupied by Fe atoms in the unit cell, as shown in Fig. 3.26a. In an ideal inverse $CuFe_2O_4$ spinel structure, the tetrahedral hollow sites are occupied by Fe atoms while the octahedral hollow sites are occupied by equal Cu and Fe atoms, as shown in Fig. 3.26b. To be mentioned, a large supercell should be used with the random occupation of Fe and Cu atoms at octahedral hollow sites to simulate the real structure of inverse spinel $CuFe_2O_4$ [11]. Both in the normal and inverse spinel structures, O atoms are positioned in the center of a tetrahedral structure with four neighbors of Fe or Cu atoms.

In Table 3.7, the optimized lattice parameters and average binding energy per atom of normal and inverse $CuFe_2O_4$ spinel structures based on DFT and DFT + U methods are summarized. The average binding energy is defined as $E_b = (E_{total}\text{-}n_{Cu}E_{Cu}\text{-}n_{Fe}E_{Fe}\text{-}n_OE_O)/(n_{Cu} + n_{Fe} + n_O)$, where E_{total} is the total energy of $CuFe_2O_4$ per unit cell, E_{atom} is the total energy of the atom, and n is the number of atoms within the unit cell. It can be found DFT + U predicted little larger lattice constants than DFT method. The calculated binding energy suggested the inverse spinel structure is more stable than the normal one with DFT + U method, while DFT method gives the contrary result. Thus, only DFT + U method was adopted to perform the following studies.

In Fig. 3.27, the electronic band structures of the inverse spinel structure of $CuFe_2O_4$ bulk was plotted with spin-polarized DFT + U method. The inverse spinel $CuFe_2O_4$ is a spin-polarized semiconductor with an indirect band of 1.429 eV. The profile of spin-charge density, the total density of states (DOS) and partial DOS (PDOS) projected on Cu, Fe, and O atoms are plotted in Fig. 3.28. From the plotted spin-charge density distribution, the Fe atoms located at A and B sites are coupled with the antiferromagnetic method. The local magnetic moments of Fe atoms at A and B sites are −4.08 and 4.17 μ_B, respectively. Cu atoms and Fe

Table 3.7 Lattice constants, average binding energies of spinel, and inverse spinel structures of $CuFe_2O_4$ based on DFT and DFT + U methods, respectively

	Method	Spinel	Inverse spinel
a, b, c	DFT	5.777, 5.777, 5.777	5.976, 5.976, 5.676
	DFT + U	5.917, 5.917, 5.917	6.059, 6.059, 5.912
α, β, γ	DFT	60.00, 60.00, 60.00	61.64, 61.64, 57.03
	DFT + U	60.00, 60.00, 60.00	60.80, 60.80, 57.57
$E_{binding}$ (eV)	DFT	4.704	4.630
	DFT + U	3.719	3.992

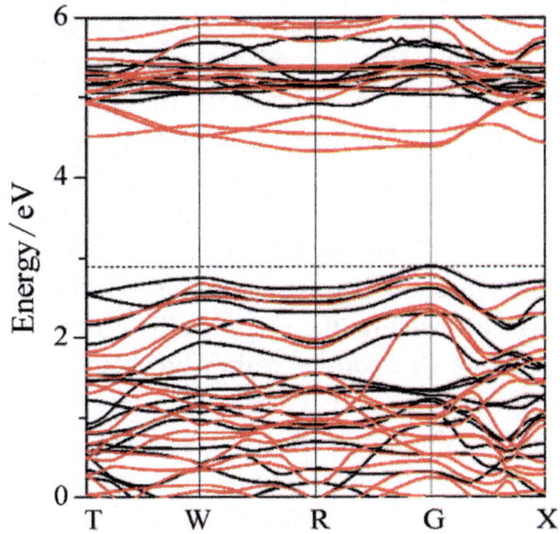

Fig. 3.27 Electronic band structures of the inverse spinel $CuFe_2O_4$ with spin-polarized DFT + U method [10]

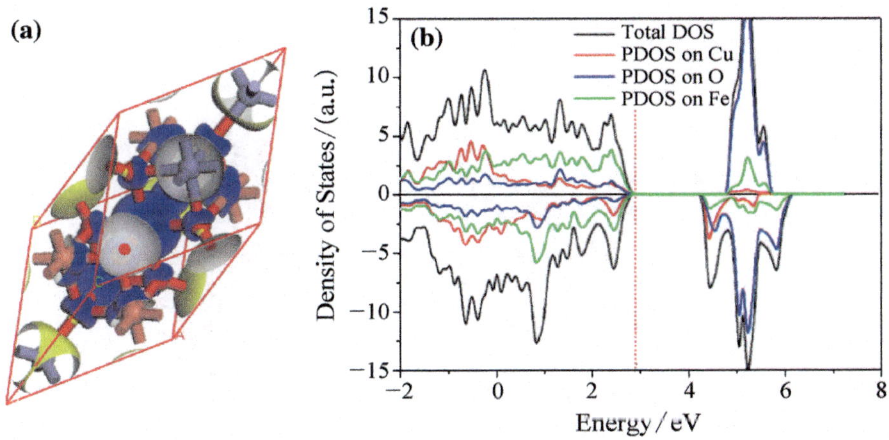

Fig. 3.28 **a** Iso-value surface of the spin electronic density distribution of inverse spinel $CuFe_2O_4$ in the unit cell [10]. The blue and yellow represent ±0.03 a.u. **b** The total DOS and partial DOS for the major and minor spin of $CuFe_2O_4$. The Fermi level was plotted with the red dotted line

atoms at B sites are coupled in the ferromagnetic method, and the local magnetic moment of Cu atom is 0.62 μ_B. The magnetic order consistent with previous work [12], where the magnetic moments of ions within the same sites (A or B site) are ferromagnetically ordered, and those between different sites (A and B sites) are antiferromagnetically ordered O atoms have a weak magnetic moment of 0.068 μ_B. The total magnetic moment of $CuFe_2O_4$ is 1.98 μ_B per unit cell. From the calculated DOS and PDOS, the occupied states just below the Fermi level are mainly contributed by the coupling of d electrons from Cu and Fe atoms and p electrons from O atoms, while the unoccupied states just above the Fermi level are mainly contributed by p orbitals of O atoms.

The major and minor spin bands are plotted with black and red lines, respectively. The Fermi energy level was plotted with a dotted line. *T*, *W*, *R*, *G*, and *X* represent (0.5, −0.5, 0.0), (0.75, −0.25, −0.25), (0.5, 0.0, 0.0), and (0.5, 0.5, −0.5) *k*-point in the first Brillouin zone, respectively.

As shown in Fig. 3.29, we built 2 × 2 (100) surface from the inverse spinel structure of $CuFe_2O_4$ to study the adsorption behavior of NO molecule on the surface. Thirteen atomic layers plus vacuum layer of 20 Å thickness were chosen to simulate the half-infinite structure of the surface. When performing optimization of surface structure, the atoms in the bottom seven atomic layers are fixed at their bulk positions (the atoms within the black box, as shown in Fig. 3.29a), while other atoms are relaxed without any constraint. The magnetic orders of the atoms are same as those inside the perfect bulk. After the geometric relaxation, the surface atoms have reconstruction compared with their original bulk positions, as shown in Fig. 3.29b. Compared with the bulk structure, the surface O atoms shift up 0.82 Å, while surface Cu and Fe atoms shift down 0.54 Å. The sixth atomic layer

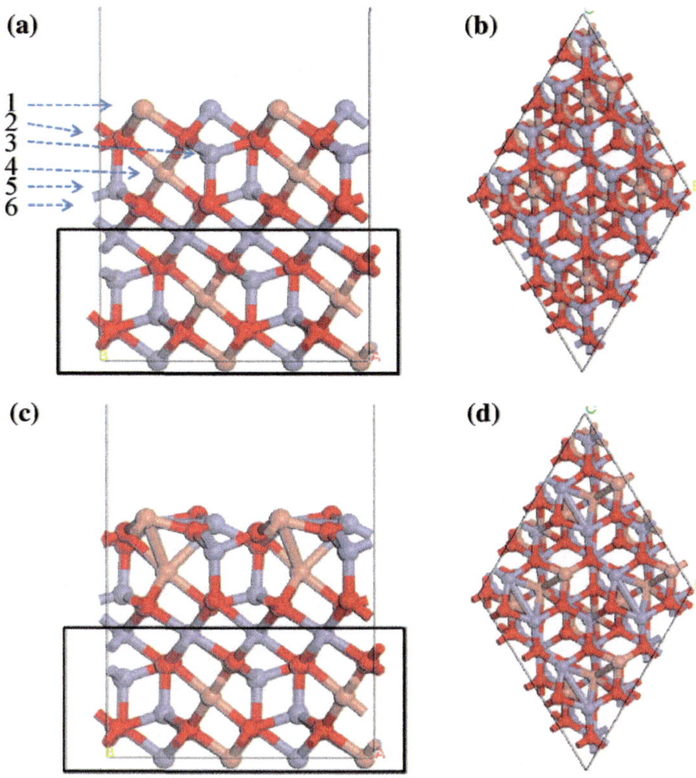

Fig. 3.29 **a**, **b** Are the front and top views of 2 × 2 (100) surface of the $CuFe_2O_4$ inverse structure before the atomic relaxation, respectively; **c**, **d** are the front and top views of 2 × 2 (100) surface of the $CuFe_2O_4$ inverse structure after the atomic relaxation, respectively [10]

accounting from the top layer has a very tiny vertical shift of about 0.06 Å, indicating the atomic positions in this atomic layer are very close to their bulk positions. The total magnetic moment is about 32.41 μ_B.

The adsorption of NO molecule was studied on the reconstructed $CuFe_2O_4$ surface. Four initial structures were considered, including (1) NO molecule was adsorbed on the top of surface Cu atom with the N atom of the molecule toward the Cu atom (ON–Cu), (2) NO molecule was adsorbed on the top of surface of Cu atom with the O atom of the molecule toward the Cu atom (NO–Cu), (3) NO molecule was adsorbed on the surface Fe atom with the N atom toward the Fe atom (ON–Fe), and (4) NO molecule was adsorbed on the surface Fe atom with the O atom toward the Fe atom (NO–Fe). All atoms in the surface have the same initial magnetic order as those in bulk when performing geometric optimization.

The optimized structures of four situations are presented in Fig. 3.30, and the adsorption energies of NO molecule on the surface, the charge distribution over the NO molecule, as well as the geometric parameters of the four optimized structures are summarized in Table 3.8. The adsorption energy is defined as $E_{ads} = E(CuFe_2O_4 + NO) - E(CuFe_2O_4) - E(NO)$, where E(system) is the total energy of the corresponding system. From the calculated adsorption energy, all configurations have negative values of adsorption energy, suggesting that the adsorption of NO molecule on $CuFe_2O_4$ (100) surface is exothermic. NO molecule prefers to adsorb on the surface Cu or Fe atom with N atom connected to the surface atom. The most favorable configuration is that NO molecule adsorbs over the surface Fe atom (NO–Fe), forming an N–Fe bond. The calculated adsorption energy is −1.98 eV. The adsorption of NO molecule over the surface Fe atom with the formed O–Cu bond

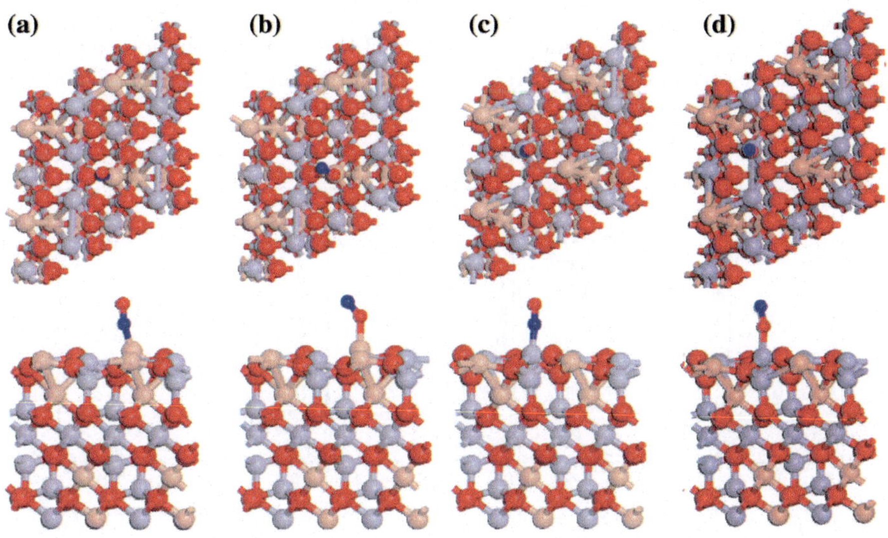

Fig. 3.30 Optimized configurations of NO adsorption on the$CuFe_2O_4$ surface, **a** ON–Cu, **b** NO–Cu, **c** ON–Fe, **d** NO–Fe [10]

Table 3.8 Calculated adsorption energy of NO molecule on the surface, equilibrium N–M or O–M (M=Cu or Fe) bond length, N–O bond length within the NO molecule, the charge distribution over the NO molecule, and the O–N–M or N–O–M bond angles [10]

	ON–Cu	NO–Cu	ON–Fe	NO–Fe
E_{ads} (eV)	−1.05	−0.53	−1.98	−0.33
N–M or O–M (Å)	1.76	1.93	1.79	1.86
N–O (Å)	1.19	1.21	1.19	1.21
$Charge_{NO}$ (e)	1.70	1.66	1.60	2.43
O–N–M (°)	149.31	/	166.74	/
N–O–M (°)	/	135.21	/	170.04

(NO–Fe) has the smallest adsorption energy of about −0.33 eV. In the ON–Fe configuration, the N–Fe bond length is 1.78 Å. Due to the interaction between the NO molecule and $CuFe_2O_4$ surface, about 1.60*e* charge is transferred from NO molecule to the surface, resulting in the weak of the N–O bond within the molecule. The charge analysis result is obtained with worse method [13]. The N–O bond length is elongated to 1.19 Å, which is longer than that of the N–O bond in the free NO molecule (1.17 Å at the same calculation level). The total magnetic moment of NO–Fe configuration is about 31.96 μ_B, which is a little smaller than that of the pure $CuFe_2O_4$ surface. In all other configurations, NO molecules are charged with positive charges with N–O bond elongated compared with that in the free NO molecule.

To understand the interaction between NO molecule and $CuFe_2O_4$ surface, the density of states projected on NO molecule and the 3d orbital of Fe atom bound to NO molecule of the ON–Fe configuration are plotted for spin-up and -down components in Fig. 3.31a, b, respectively. The σ and π-orbitals of NO molecule are defined as $\pi = N_{px} + N_{py} + O_{px} + O_{py}$, and $\sigma = N_s + N_{pz} + O_s + O_{pz}$, where A_b means the "*b*" type atomic orbital of element *A* [14]. Upon the spin-up component, there is a strong hybridization between NO $2\pi^*$ orbital and Fe 3d states. Upon the

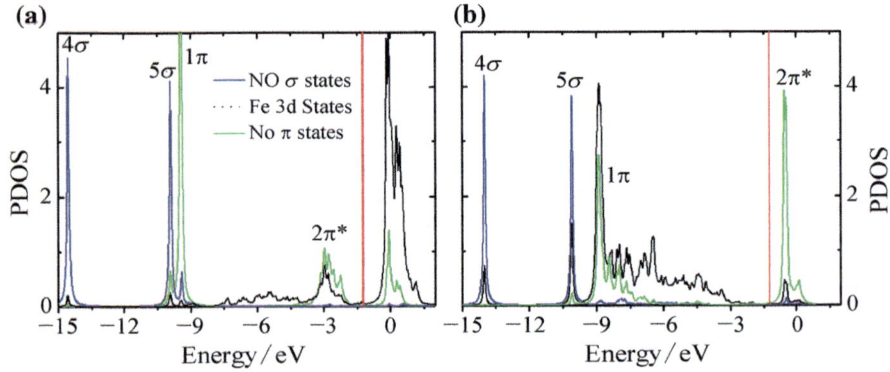

Fig. 3.31 Partial density of states (PDOS) of adsorbed NO molecule and the 3d states of directly contacted Fe atom, **a**, **b** represent spin-up and spin-down component, respectively [10]

spin-down component, there is a strong hybridization between NO 1π orbital and Fe 3d states due to the empty NO $2\pi^*$ orbital. Thus, the electron is donated from NO π orbitals to Fe 3d states. It is also noticed that the hybridization between NO σ orbital and Fe 3d states is negligible, which indicates a weak back-donation process between NO and Fe for the high oxidation state of Fe atom. Thus, the strong adsorption mostly comes from the hybridization of π orbitals, and 3d states and electrons are transferred from NO to Fe^{3+}.

In conclusion, we studied the ideal adsorption situation of NO molecule on the surface of $CuFe_2O_4$ with DFT + U method. Our studies reveal that the on-site Coulomb interaction within Cu and Fe 3d states should be considered when dealing with $CuFe_2O_4$, and the inverse spinel structure of $CuFe_2O_4$ is energetic favorable. In the inverse spinel $CuFe_2O_4$, Fe atoms occupied the tetrahedral hollow site and octahedral hollow sites equally, while Cu atoms occupied the remaining octahedral hollow sites. The Fe atoms at tetrahedral and octahedral sites are coupled in anti-ferromagnetic method with a local magnetic moment of about 4 μ_B. The Cu atoms and Fe atoms at the octahedral sites are coupled in the ferromagnetic method, and the local magnetic moment of Cu atoms is 0.62 μ_B. NO molecule prefer to adsorb on the top site of surface Fe atom with the formed N–Fe bond over the (100) surface of the inverse spinel $CuFe_2O_4$. The adsorption is exothermic with the adsorption energy of about −1.98 eV. Due to the interaction between NO molecule and the surface, about 1.60 e charges are transferred from the molecule to the surface, and the N–O bond within the molecule is elongated. And the strong adsorption mostly comes from the hybridization of π orbitals (NO) and 3d states (Fe^{3+}), and electrons are transferred from NO to Fe^{3+}, which will lead to the further NO activation/reduction reaction.

References

1. J.M. Smith, Chemical engineering kinetics (1981)
2. W. Shangguan, Y. Teraoka, S. Kagawa, Simultaneous catalytic removal of NO_x and diesel soot particulates over ternary AB_2O_4 spinel-type oxides. Appl. Catal. B **8**, 217–227 (1996)
3. P. Ciambelli, P. Corbo, P. Parrella, M. Scialò, S. Vaccaro, Catalytic oxidation of soot from diesel exhaust gases: 1. Screening of metal oxide catalysts by TG-DTG-DTA analysis. Thermochim. Acta **162**, 83–89 (1990)
4. P. Ciambelli, M. D'AMORE, V. Palma, S. Vaccaro, Catalytic combustion of methane soot. Combust. Sci. Technol. **121**, 67–84 (1996)
5. X. Chu, L.D. Schmidt, Intrinsic rates of nitrogen oxide (NO_x)-carbon reactions. Ind. Eng. Chem. Res. **32**, 1359–1366 (1993)
6. W. Shangguan, Y. Teraoka, S. Kagawa, Kinetics of soot—O_2, soot—NO and soot—O_2—NO reactions over spinel-type $CuFe_2O_4$ catalyst. Appl. Catal. B **12**, 237–247 (1997)
7. Y. Teraoka, W. Shangguan, S. Kagawa, Reaction mechanism of simultaneous catalytic removal of NO_x and diesel soot particulates. Res. Chem. Intermed. **26**, 201–206 (2000)
8. A. Raj, Z. Zainuddin, M. Sander, M. Kraft, A mechanistic study on the simultaneous elimination of soot and nitric oxide from engine exhaust. Carbon **49**, 1516–1531 (2011)

9. C. Garcia-Fernandez, S. Picaud, M.T. Rayez, J.C. Rayez, J. Rubayo-Soneira, First-principles study of the interaction between NO and large carbonaceous clusters modeling the soot surface. J. Phys. Chem. A **118**, 1443–1450 (2014)
10. Z. Jiang, W. Zhang, W. Shangguan, X. Wu, Y. Teraoka, Adsorption of NO molecule on spinel-type $CuFe_2O_4$ surface: a first-principles study. J. Phys. Chem. C **115**, 13035–13040 (2011)
11. S.H. Wei, S.B. Zhang, First-principles study of cation distribution in eighteen closed-shell $A_{II}B_{III2}O_4$ and $A_{IV}B_{II2}O_4$ spinel oxides. Phys. Rev. B **63**, 33–41 (2001)
12. G.F. Goya, H.R. Rechenberg, J.Z. Jiang, Structural and magnetic properties of ball milled copper ferrite. J. Appl. Phys. **84**, 1101–1108 (1998)
13. G. Henkelman, A. Arnaldsson, H. Jónsson, A fast and robust algorithm for Bader decomposition of charge density. Comput. Mater. Sci. **36**, 354–360 (2006)
14. Z.H. Zeng, S.J. Da, W.X. Li, Theory of nitride oxide adsorption on transition metal (111) surfaces: a first-principles investigation. Phys. Chem. Chem. Phys. **12**, 2459–2470 (2010)

Chapter 4
Influencing Factors for Simultaneous NO_x–Soot Removal

Apart from the material and kinetics, other several factors are affecting the multiplex reaction, simultaneous NO_x–soot removal via catalysts. It is believed that most catalytic reactions occur on the surfaces. Hence, the discussions on surface modification in addition to composition design are indispensable to this process. The surface modifications by a variety of preparation techniques will consequently cause the alteration of many macrolevel parameters including surface areas and morphology and in turn influence the reactant/intermediate adsorption and desorption behaviors on the surface. The interaction between the components of catalysts, for instance, metal oxides and noble metals, plays a crucial role in this reaction. However, its inherent mechanism is still unclear and under debating. In several specific atmospheres, some catalyst behaves contrary to common knowledge, which is worthy of further discussion and investigation. Also, the interaction between catalyst and soot is also to be evaluated since the reaction rate is partially determined by contact conditions between the two solids.

4.1 Contact Conditions

In a solid–solid–gas reaction system, the interaction between gas and solid is realized by the conventional theory below. Gas experiences the following steps for a complete catalytic reaction.

(1) Diffusion of reactants into the boundary layer and internal pores (internal diffusion).
(2) Adsorption on the catalysts.
(3) Dissociation of adsorbed gasses and the formation of products.
(4) Desorption from catalysts surface.
(5) Diffusion of reactants into the boundary layer and internal pores (external diffusion).

W. Shangguan et al., *Simultaneous Catalytic Removal of Diesel Soot and* NO_x,
Energy and Environment Research in China,
https://doi.org/10.1007/978-981-10-7266-6_4

The gas–solid reaction itself is a complicated process judging by these steps, namely the interaction between the gas and solid, whereas solid–solid–gas reaction is more pronounced to be discussed since the interaction between soot and catalyst is also a preferentially considered factor. Many physical parameters including as heat transfer, mass transfer, the contact intensity, and the contact area are so hard to trace during the reaction. Thus, a "black box" approach is adopted to investigate the contact condition; that is, to focus on the correlations between experimental results and initial conditions with the neglect of intermediate process. In terms of the homogeneity of the mixture, a proper soot/catalyst ratio is to be picked. On the basis of this methodology, the debate of interaction between the two solids can be further categorized into two states: static state and dynamic state.

4.1.1 Static State

The phenomenon that contact conditions significantly alter the reaction behavior was pioneered in 1992 by van Doorn [1] and further evidenced by Mul in 1995 [2]. However, these results are tough to reproduce. In 1998, Teraoka utilized modeled soot to prepare a soot/catalyst mixture, which is denoted tight contact condition, to replicate the contact status [3]. On the other hand, Neeft discovered that the "loose contact" state resembles the working contact condition [4]. Regarding soot–NO–O_2 reaction, the enhancement of soot–catalyst contact accelerates soot consumption, and in turn, the NO_x reduction. It is presumed that the reaction dominantly takes place at the three-phase boundary. It is essential to study soot–O_2 reaction, a simplified model of soot–NO–O_2, to dig into the complex contact condition between soot and catalyst.

In the last decades, various contact modes have been created via exerting mechanical into soot/catalyst mixture, which is summarized in Table 4.1. Super tight contact mode is recently visualized by Aneggi group [5]. In this study, cerium–zirconium solid solution materials were synthesized and high-energy ball milling of soot/catalyst mixture for various times was acquired. The morphology of selected samples before and after milling was investigated by high-resolution

Table 4.1 Summary of preparation method and variable for various contact mode

Contact mode	Preparation method	Variable
Super tight contact	High-energy ball milling	Time
Pressure contact	Pressurization via a preformation machine	Pressure and time
Tight contact	Grind in a mortar	Time
Wet contact	Solution agitation along with evaporation	Solution type and evaporation temperature
Loose contact	Mixed by spatula	Time

transmission electron microscopy (HRTEM). Figure 4.1a shows an HRTEM image of a $Ce_{0.8}Zr_{0.2}O_2$ sample and soot mixed under tight contact mode. The sizes of oxide nanoparticles range between 4 and 6 nm. The Fourier Transform (FT) image shows rings corresponding to the fluorite planes characteristic of $Ce_{0.8}Zr_{0.2}O_2$ (CZ). Figure 4.1b displays the spots of CZ as well as wide patches at 3.4 Å, which are characteristic of soot. After milling for 1 h, the distribution of the particle size is much broader and shifted toward larger sizes. In Fig. 4.1c, large particles of CZ (10–15 nm) are visible along with soot globules. The 4 h milling causes the disappearance of the majority of soot aggregates, and soot can be visualized prevalently as a thin shell over the CZ crystallites (Fig. 4.2a). After 8 h milling, there is no soot clumps and soot has formed a thin layer of the ceria particles in intimate contact (core-shell structure). In the HRTEM image depicted in Fig. 4.2b, a thin layer of soot is recognized over the ceria crystallites. This soot layer is even more

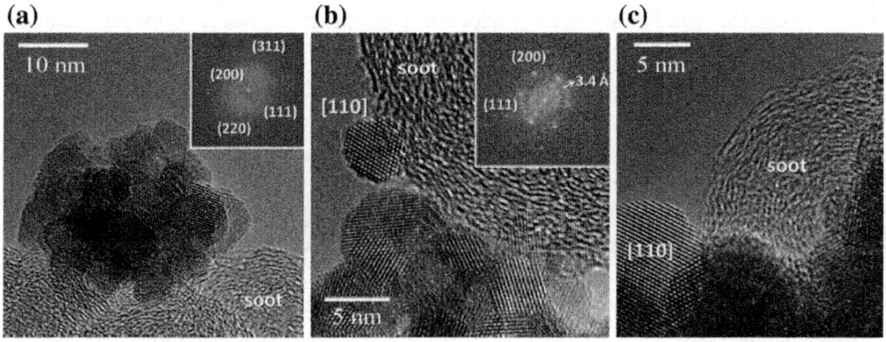

Fig. 4.1 HRTEM images of $Ce_{0.8}Zr_{0.2}O_2$ and soot mixed, **a**, **b** in tight contact mode (weight ratio catalyst: soot of 20:1), and **c** after milling for 1 h. In all cases, soot globules (lattice fringes at 3.4 Å) are identified in contact with crystalline CZ nanoparticles [5]

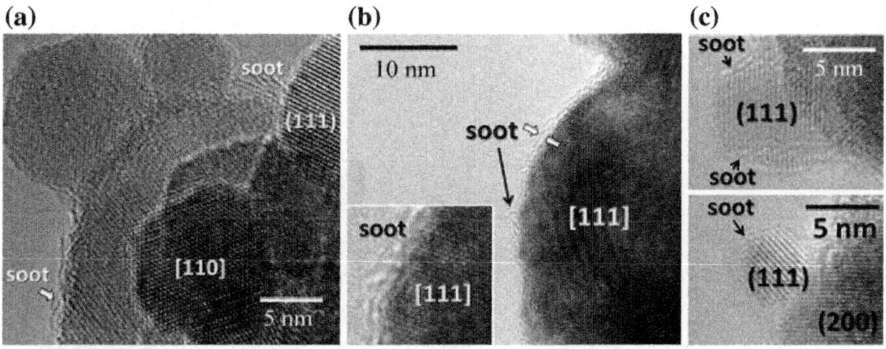

Fig. 4.2 HRTEM images of $Ce_{0.8}Zr_{0.2}O_2$ and soot (weight ratio catalyst to soot of 20:1) [5], **a** after milling for 4 h, **b** and **c** 8 h. Soot aggregates have disappeared upon milling and soot is present as a thin shell wrapping the CZ crystallites

disordered than in the previous samples. Figure 4.2c displays two ceria–zirconia nanoparticles showing lattice fringes at 3.11 Å corresponding to (111) crystallographic planes of the CZ lattice. These crystallites are inclined to be covered by a quasi-monoatomic soot layer, which should be highly reactive [5].

Figure 4.3 illustrates that the extended milling time shows a positive effect on soot combustion by increasing contact points. The "core-shell" structure of soot/catalyst mixture maximizes the contact points and the catalytic activity. In addition to this regard, Aneggi believes that the high-energy milling progressively reduces the presence of sizeable 3D soot clumps and distributing more or less uniformly the carbon over the catalyst by wetting the catalyst surface with an almost 2D thin layer of carbon. This could also be a crucial factor for soot combustion.

Lars Hensgen prepared soot/catalyst mixture with various contact conditions over a reference material (CeO_2) and three sorts of soot (PU, P90, and FS) (Fig. 4.4) [6]. Four contact modes are applied in this study. The tight contact and loose contact condition are well known in a variety of literature. Apart from this, the other two contact modes are also produced. The pressure contact condition can be

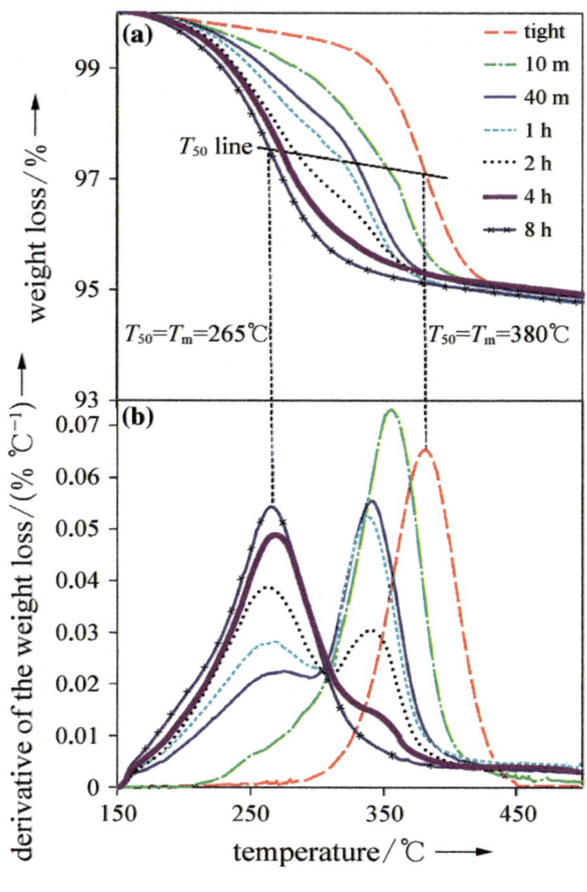

Fig. 4.3 TGA profiles in the air of CZ/C mixtures (catalyst/soot ratio of 20:1) treated at different milling time [5], **a** weight loss curve (% wt) with an indication of the temperature at which 50% of soot weight is lost under working conditions, **b** the corresponding derivative weight loss function with peak profiles because of oxidation. The limit values of T_{50} and T_m measured after 0 (tight), and 8 h milling are highlighted

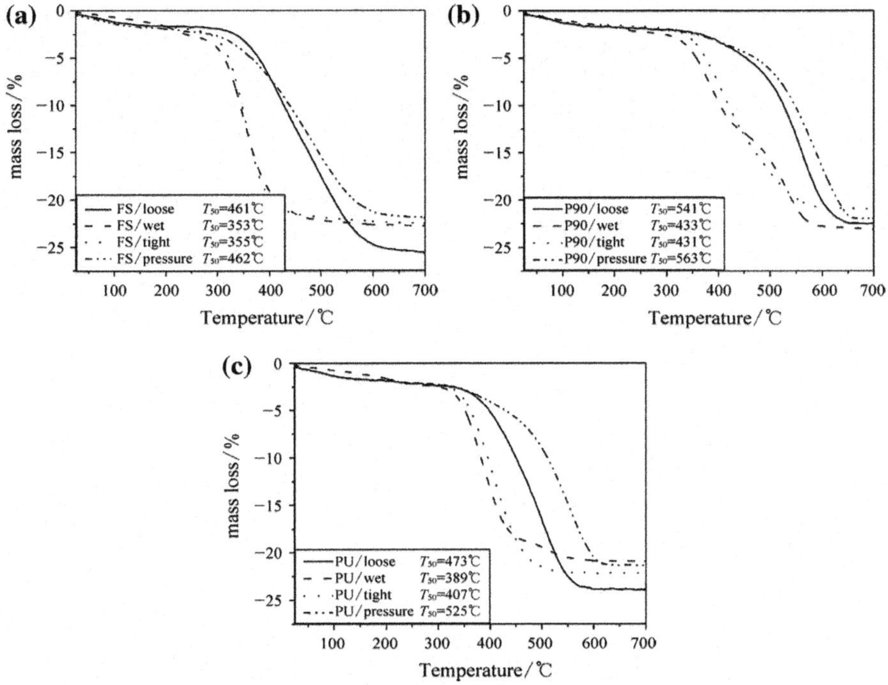

Fig. 4.4 Thermogravimetric measurement with one soot and different contacts [6], **a** fullerene soot contact, **b** Printex U soot, **c** Printex 90 soot (8% O_2 in N_2; 25–700 °C with 2 °C/min)

acquired by the following method. The mixtures are pressed to pellets employing a hydraulic press with high pressure (typically 40 MPa). The pellets are then granulated and sieved in mesh size of 250–500 μm [7]. In parallel, the wet contact mode is obtained via stirring 1 g CeO_2 and 250 mg of soot in 15 mL of an organic solvent such as acetone with 350 rpm of a magnetic stirring bar for 6 h. Subsequently, they are dried at 40 °C. In Fig. 4.4, the soot combustion temperatures in their absolute form are strongly dependent on the contact mode. For the three classic contact modes exist in a fixed order concerning the T_{50} value: tight < loose < pressure. The newly created wet contact is not located at a fixed position within this sequence. While with PU soot on VP AdNanoCeria 50, the wet contact is between the tight and the loose contact, with P90 and FS the T_{50} value is even lower than in the case of the tight contact. The wet contact mode is unique in the generation of a homogeneous mixture of soot and catalyst. In addition to this, the wet contact mode somehow resembles the working condition due to the residual SOF wrapped on soot surface, and low-temperature drying process preserves the SOF content to a maximum level. However, this property challenges the evaluation parameters; it is not proper to apply the sole T_{50} to overview the whole combustion process, since it is tough to distinguish soot and SOF consumption, respectively. The effect of different solvent, which can be eventually ascribed to the effect of residual SOF, is displayed in Table 4.2. The patterns and their intensities for heat flux are also recorded for reference. The intensity of heat-flux peaks from solution to solution which indicates the contact intensity also alters. As shown in Table 4.2, the stronger intensity of double peak structure leads to the better catalytic performance. The innovation of wet contact is enormous progress in quantifying the contact intensity to some extent (Table 4.3).

A prolonged loose contact mode was acquired over CeO_2 of distinct morphology including fibers, flakes, sticks, and SCS powders. This contact mode is obtained via

Table 4.2 Influence of different solvents on the double peak effect of the wet contact mode [6]

Solvent	Double peak structure	T_{50} (°C)
Acetone	Strong	433
Pentane	Strong	496
Isopropanol	Middle	513
Hexane	Middle	519
Methanol	Middle	521
Ethanol	Middle	522
Isooctane	Middle	539
Toluene	Weak	540
Tetrahydrofuran	Weak	544
Benzene	Weak	550
Acetonitrile	Weak	552
Chloroform	No	569
Ethyl acetate	No	579
Water	No	588

Table 4.3 Soot combustion activity results, under "loose," "prolonged loose," and "tight" contact conditions, of CeO_2 fibers, flakes, sticks, and (solution combustion synthesis) SCS-derived powders [8]

	Contact	$T_{10\%}$ (°C)	$T_{50\%}$ (°C)	T_{peak} (°C)
Printex U		474	575	600
Fibers	Tight	317	372	372
	Prolonged loose	344	426	428
	:Loose	395	533	553
Flakes	Tight	356	420	425
	Prolonged loose	392	428	475
	:Loose	450	552	567
Sticks	Tight	375	445	425
	Prolonged loose	410	498	523
	:Loose	418	537	557
SCS powders	Tight	310	365	365
	Prolonged loose	365	461	459
	:Loose	393	525	535

the following technique: The catalyst–carbon mixture was prepared by extending the mixing time, still with the same technique described for the loose contact. In this case, a better contact was established between the soot particles and the catalyst [8]. It is apparent that the prolonged loose contact mode favors soot ignition at a lower temperature than loose contact mode does, which indicates the contact points and/or contact intensity can be improved via prolonging the preparation time of soot/ catalyst mixture.

In addition to the above-mentioned techniques to prepare soot/catalyst mixture, the morphology of catalyst is likely to exert a significant effect on the catalytic performance. A variety of papers focus on macroporous material, such as 3D material, and it is believed that the macropores can "accommodate" soot and enhance the contact efficiency between soot and catalyst [9–12]. This is absolutely true. However, in this case, it is hard to exclude of the contribution of exposure of active crystal face in such studies and heat/mass transfer process is still a thorny issue.

To conclude, the static contact mode can be classified into the following three circumstances which are listed in Table 4.4. The face-to-face contact model usually tallies with pressure contact and super tight contact condition while the latter two (point-to-point contact and no contact condition) resembles the practical condition, namely loose contact. Another issue to consider is contact intensity which involves the mechanical energy exerted by the contact modes.

4.1.2 Dynamic State

The visualization of the soot oxidation via environmental transmission electron microscope (ETEM) is a hot spot over recent years. The technique provides the

Table 4.4 Three modes in static contact condition

Contact mode	Diagrammatic sketch	Performance
Face-to-face contact	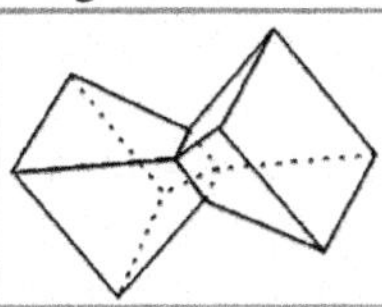	High
Point-to-point contact	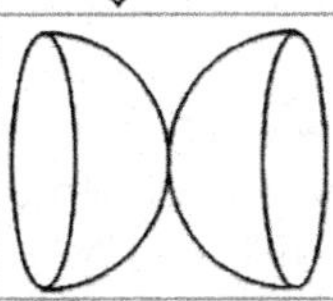	Medium
No contact	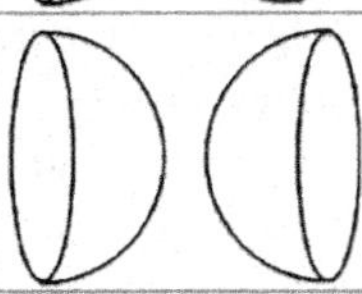	Low

possibility to trace the motion of soot and catalyst particle, respectively. In fact, it is a useful tool to observe the details of contact between soot and catalyst.

The application of ETEM in the field of soot combustion starts since 2008 [13]. With the aim of addressing the role of the CeO_2, ETEM experiments were performed to monitor agglomerates of carbon black (CB) particles attached to CeO_2 particles in situ during exposure to oxygen at elevated temperatures. The time-lapsed series of ETEM images show the main findings: The images reveal that agglomerates of CB particles "disperse" the CeO_2 and vanishes (Fig. 4.5a–c), whereas no eminence is discovered over the agglomerates of CB particles locates at the CeO_2 particles. These findings suggest that the motion is ascribed to not only to the migration of CB along the CeO_2 surface but also to the oxidizing reaction conditions. This finding is consistent with the results of similar experiments using N_2 instead of O_2, in which ETEM image series shows that CB particle agglomerates remained stable. Moreover, monitoring the CB–Al_2O_3 samples by ETEM during exposure to the oxidizing conditions (Fig. 4.5d–f) shows that the agglomerates of particles are stable at the same temperature and over the same time period. Thus, the findings imply that the CeO_2 catalyzed the oxidation of the CB agglomerates and concurrently generated motion of the agglomerates toward the catalyst surface [13]. In this case, the motion of soot is observed, and the contact condition is not promoted via the attack of catalysts. Similar findings are acquired over the silicon-oxy-carbide layer (with Pt) [14]; soot collapse at the boundary of catalyst surface with the elapse of time. As mentioned above, the motion of carbon is irrespective of the catalysts; hence, the motion of catalytic material is more decisive for the overall contact condition between soot and catalyst (Fig. 4.6).

Silver is a newly developed catalyst for soot combustion, yet its catalytic mechanism on soot elimination is still controversial. The mobility is one of the contributors to soot oxidation, which is addressed by Gardini et al. [15]. The soot combustion over silver particles under loose contact condition is recorded in Fig. 4.7. In the temperature range, 25–280 °C soot or Ag morphology remains unchanged (Fig. 4.7a). As the temperature elevates from 280 to 472 °C, silver particles begin to coalesce, forming more agglomerates (Fig. 4.7b). At 500 °C coalesced silver agglomerates are ready to be mobile, moving on the soot cake, and actively oxidizing the soot particles. Soot oxidation at the Ag/C interface is visually confirmed by the disappearance of soot particles in contact with silver. A similar case occurs for Ag/soot in tight contact conditions that the mixture experiences three primary stages for the whole reaction process: silver particle coalescence, the mobility of coalesced silver agglomerates, and massive soot consumption due to the attack from the motional silver agglomerates. More microscopic scale information for loose contact mixture was acquired BF-TEM micrographs (Fig. 4.9). In loose contact condition, silver particles in larger clusters are kept together by the Ag–Ag cohesive energy from large agglomerates in touch with the soot cake (Fig. 4.9a). Upon heating process, silver particles tend to coalesce forming round agglomerates (Fig. 4.9b). Silver was observed to maintain its crystalline state after coalescence phase and throughout the rest of the in situ oxidation experiment for both contact modes. As the temperature rises, silver layers situated at the edge of the coalesced

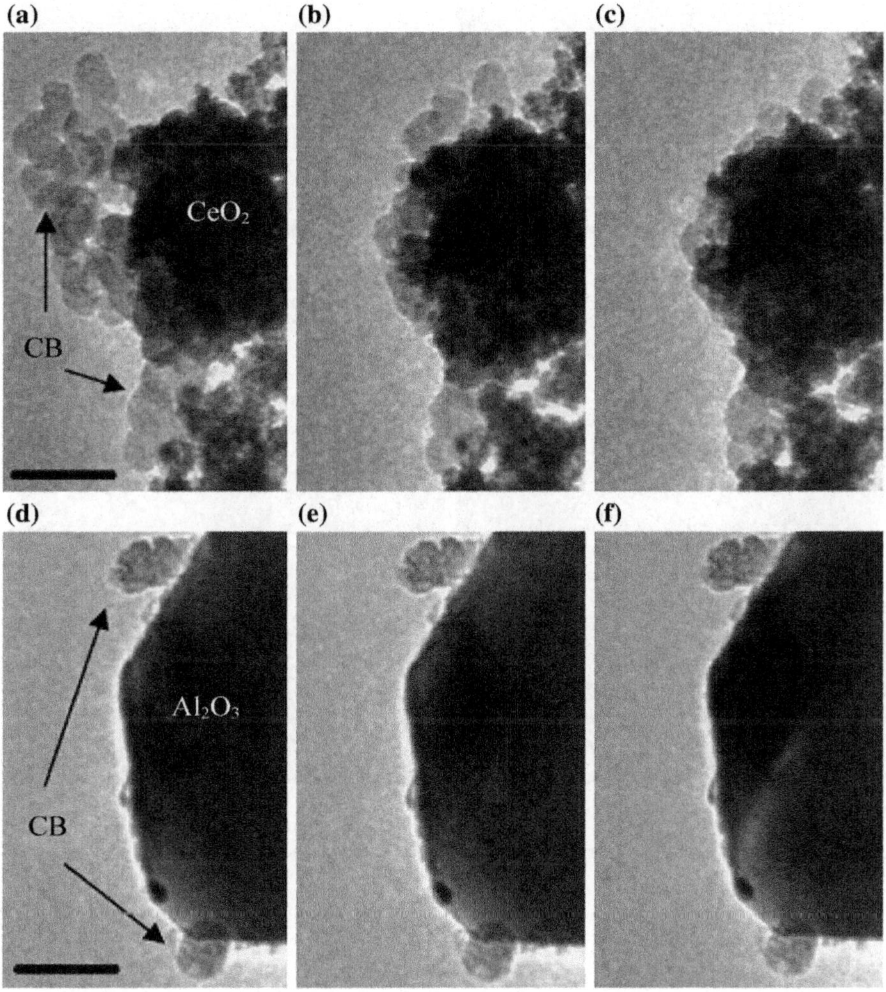

Fig. 4.5 Time-lapsed ETEM images of **a–c** soot–CeO_2, and **d–f** soot–Al_2O_3 recorded during exposure to 2 mbar O_2 at 550 °C [13]. The time interval between each image is ~2 min. Scale bars = 90 nm. Reprinted from literature

agglomerates possess sufficient energy to overcome the internal Ag–Ag cohesive energy and start wetting the soot surface causing a local deformation of the agglomerate (Fig. 4.9c). As the temperature rises to the desired value, catalytic carbon oxidation occurs; the attractive forces between silver and soot will maintain the contact between silver and the progressing oxidation front, causing a net movement of the Ag agglomerate (Fig. 4.9d) [15]. Figure 4.10 also gives another instance for the visualization of the movement of silver particles over YSZ support [16]. The silver particle is exceptionally active in mobility so that it can catalyze soot oxidation via its movement. All these reports reveal that the high mobility of

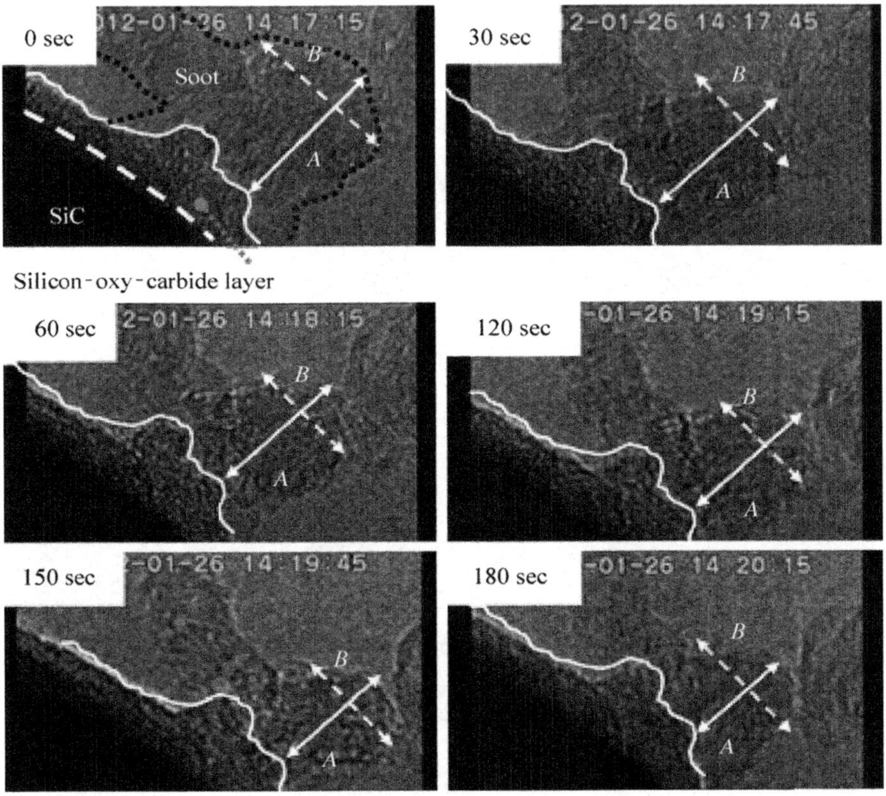

Fig. 4.6 ETEM images of soot oxidation reaction on SiC nanoparticle: DPMF (with Pt) [14] Temperature: 450 °C, Atmospheric pressure: 2.0×10^{-2} Pa, injected gas: air

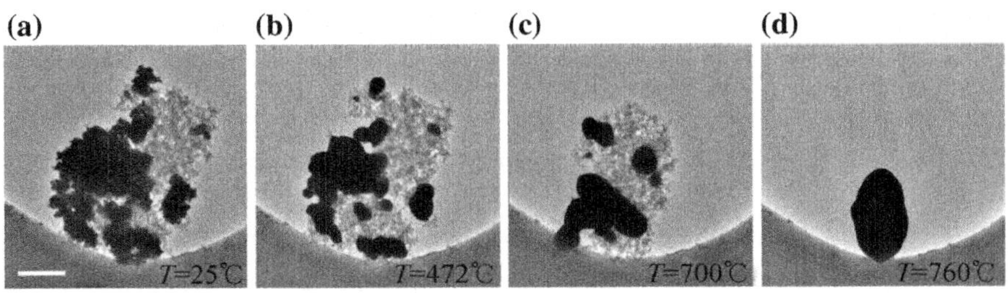

Fig. 4.7 BF-TEM micrographs of in situ soot oxidation in the ETEM in loose contact condition showing, **a** initial distribution and morphology of silver and soot, **b** coalescence of silver particles, **c** mobility of coalesced silver agglomerates over the soot cake and **d** final silver agglomeration [15] (scale bar is 500 nm)

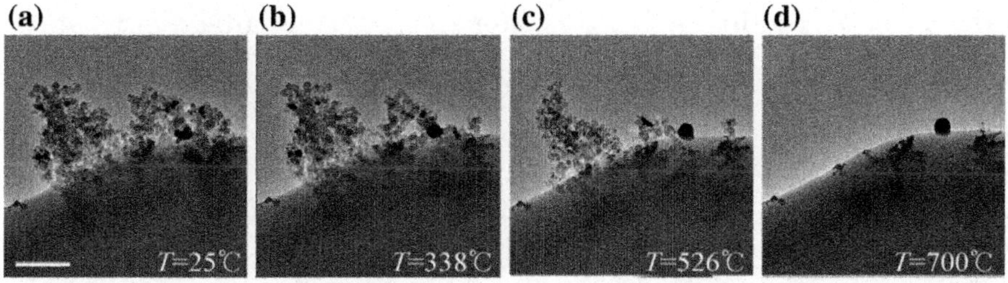

Fig. 4.8 BF-TEM micrographs of in situ soot oxidation in the ETEM in tight contact condition showing, **a** initial distribution and morphology of silver and soot, **b** coalescence of silver particles, **c** mobility of coalesced silver agglomerates over the soot cake and **d** final silver agglomeration [15] (scale bar is 300 nm)

silver shown here is of importance to the understanding of silver catalysts used for soot oxidation and should also be of relevance for other soot oxidation catalysts that could derive mobility from, e.g., a low melting point metal (Fig. 4.8).

ETEM technique can provide relatively direct evidence to study the mobility of the particles. However, it is not widely accessible. Other conventional techniques such as electron paramagnetic resonance (EPR) also provide possibilities to obtain indirect proofs for the variation of contact conditions. Here is a practice conducted by our group (Fig. 4.11). The curve is made up of two overlapped signals S1 and S2: S1 can be ascribed to soot information, and its intensity depends on the number of radicals on the soot surface; S2 reflects the contact points between soot and catalysts. In most cases, the intensity of S2 is hardly observed; hence, the contact conditions are judged by the intensity of superposition signals. The tuning of pretreatment temperature modifies the amount of volatiles on soot surface. The higher temperature leads to the release of lower radicals. The signal of Co_3O_4/soot mixture gradually decayed with elevated temperature in the inert gas. On the other hand, the signal of silver-based samples experiences a sudden uptake at 200 °C, which reveals the contribution of the S2 signal. Above 200 °C, the damping signal indicates that the increase in S2 is unable to compensate the loss of S1 [17]. At higher temperature, agglomerated silver particles tended to move around and reconstructed the contact points. The metal particles are proved to show different mobility over various substrates [18].

4.2 Precious Metal Loading

In most cases, Platinum (Pt) and Silver (Ag) are the commonly used PGM in soot combustion and simultaneous NO_x–soot removal process. It is reported that Pt can lower the ignition temperature of soot over many substrates including a metal oxide, rare earth oxides, and porous structure material [20–23]. It is recognized that Pt catalyzes the NO conversion into NO_2, thus assisting soot combustion [24–26].

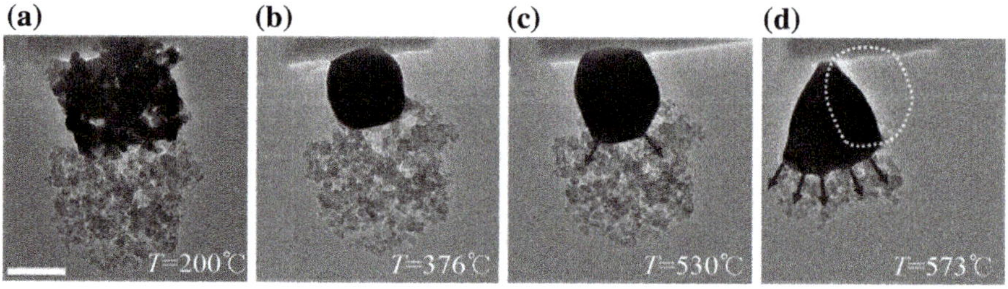

Fig. 4.9 Wetting and movement of a silver agglomerate during in situ oxidation of soot: a silver mixture in loose contact condition [15]. The previous position of the silver agglomerate as observed in (**c**) is highlighted in subfigure (**d**) with a white dashed line. Arrows in red indicate the direction of deformation of the silver agglomerate (scale bar is 200 nm)

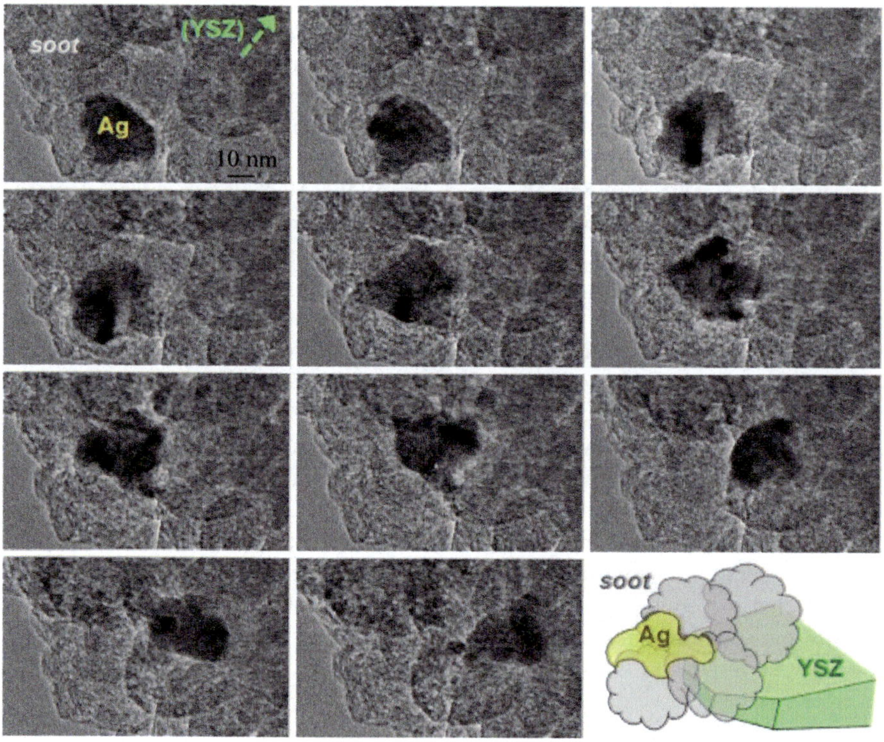

Fig. 4.10 300 kV ETEM observations of 1% Ag–YSZ + soot mixture under 495 °C, 0.5 mbar of O_2 pressure: frames extracted every 3″ from a video of about 30″ [16] (the sketch on the bottom right illustrates how the Ag NP followed in the movie deforms to adapt its shape to the topography of the soot particles)

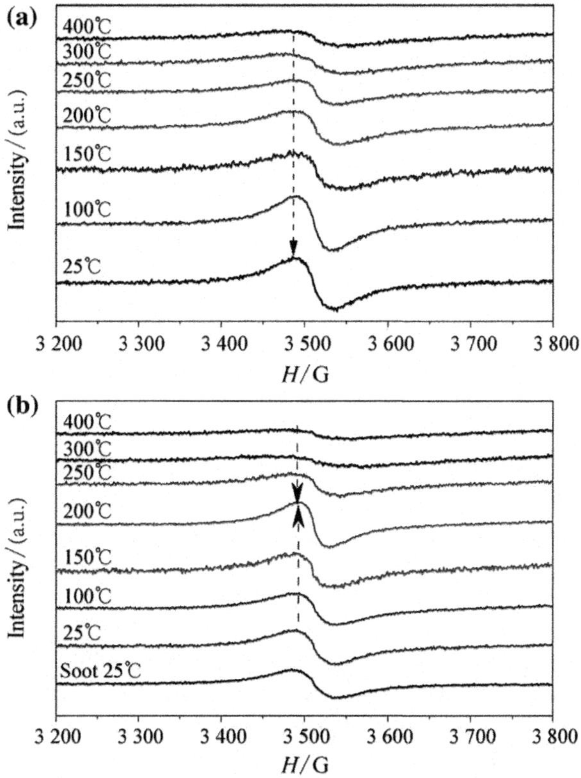

Fig. 4.11 EPR study at room temperature for mixtures pretreated in various temperatures in N_2 [19], **a** Co_3O_4 + soot; **b** 0.2Ag/Co_3O_4 + soot

Figure 2.7 reveals that the Pt loading can slightly assist soot ignition, whereas the N_2 conversion is inhibited. Pt is usually adopted as a strong oxidant. The participation of Pt in the reaction is possibly adverse to NO_x reduction; its temperature window is partially overlapped by that of soot oxidation, which indicates the oxidation reaction is more selective than the reduction reaction over the most Pt-loaded catalyst.

In terms of silver-based catalysts, they are recently reported to be extremely active for soot combustion. They have been loaded on a variety of supports such as sepiolite [27], CeO_2 [28–30], SnO_2 [31], Al_2O_3 [29, 32], TiO_2 [29]. In our recent study, we discover the anomalous performance over Ag–Co composite oxides in NO_x and O_2 atmosphere, respectively, and it is detailed presented in Fig. 4.12.

The catalytic performance of xAg/$Co_{0.93}Ce_{0.07}$ synthesized via CIT method in the 5% O_2 and NO_x atmosphere (5% O_2 + 2000 ppm NO) are displayed in Fig. 4.12. It is evident that silver-free samples ignited soot at a higher temperature in O_2 condition than the temperature in NO_x; nonetheless, the catalytic performance of silver containing samples exhibits a total opposite tendency in the two atmospheres. The light-off temperature even achieves at 197 °C over the sample with the maximum silver loading ($x = 0.3$). The light-off temperature shows a descending

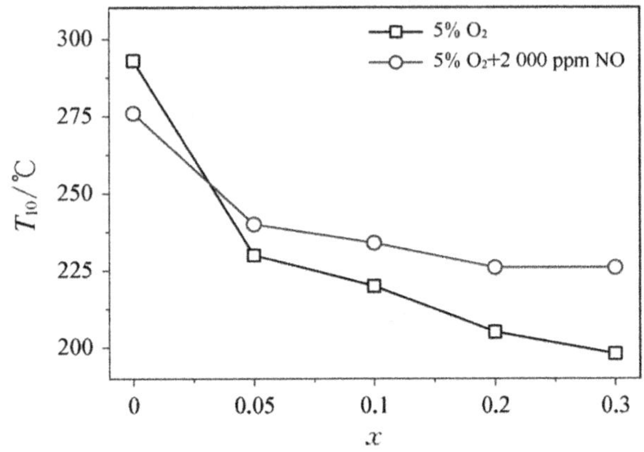

Fig. 4.12 Catalytic performances (T_{10}) over xAg/$Co_{0.93}Ce_{0.07}$ (x = 0, 0.05, 0.1, 0.2, 0.3) in 5% O_2 and 5% O_2 + 2000 ppmNO [19]

tendency with the increase of silver content in both O_2 and NO_x. Notably, both 0.2Ag/$Co_{0.93}Ce_{0.07}$ and 0.3Ag/$Co_{0.93}Ce_{0.07}$ start soot combustion at 226 °C in NO_x. These results indicate that silver plays an untraditional role in soot combustion.

For catalyst characterization, XRD patterns detect metallic silver, cubic structure Co_3O_4, and fluorite-type CeO_2; together with Raman spectra, they reveal that silver does not modify the lattice structure of Co–Ce mixed oxides. H_2–TPR illustrates that silver is incapable of boosting redox ability of Co–Ce mixed oxides. Particle size is calculated from XRD patterns ranging from 51.5 to 55.0 nm. The tiny gap of silver particle size eliminates the close relationship between this factor and catalytic performances. Besides, the BET surface area is quite small, indicating the material sintered during calcination process, and it unlikely affects the catalytic performance since this reaction took place at the three-phase boundary (soot, catalyst, and gas) [33, 34].

The summarized XPS data are displayed in Table 4.5 and Fig. 4.13. O 1 s spectra are composed of three overlapped peaks for all the materials. The peak (O_I) at 529.6 eV is ascribed to the lattice oxygen in the surface, and the shoulder signal (O_{II}) at 531.7 eV is responsible for the O^- and/or –OH) [35, 36]. It can be interpreted that O_{II} is categorized into the chemisorbed oxygen. O_{III} ($\sim$532.7 eV) peak referred to surface oxygen with weak bonds [37]. More detailed information was acquired in Table 4.4. The oxygen adsorption ability is enhanced by silver containing samples. The adsorption oxygen percentage (A_o) increased by 6.60 and 8.98% over 0.2Ag/Co_3O_4 and 0.2Ag/$Co_{0.93}Ce_{0.07}$, respectively. Moreover, CeO_2 was beneficial to the A_o value, which results from its higher oxygen ratio [38]. Soot, acting as a reducing agent, significantly alters the distribution of oxygen species on the surface. The fall of lattice oxygen ratio indicated the transformation of surface lattice oxygen into adsorption oxygen. Ag 3d profiles are composed of the Ag $3d_{3/2}$ spectrum and the Ag $3d_{5/2}$ spectrum. The latter was adopted to analyze silver surface state due to its stronger intensity. The asymmetric peak revealed the existence of silver ions in lean burn conditions. The overlapped peak consists of two peaks which refer to the metallic silver at 368 eV and the oxidized silver ($Ag^{\delta+}$) at

367 eV, respectively [39, 40]. The $Ag^{\delta+}$ concentration is higher over $0.2Ag/Co_3O_4$ than that over $0.2Ag/Co_{0.93}Ce_{0.07}$ probably because the silver oxides are partially covered by metallic silver [29]. The amount of $Ag^{\delta+}$ was not the only factor for soot combustion; the activity of the support also plays an important role in this process. The synergistic effect between CeO_2 and Co_3O_4 accelerates soot combustion, and this effect still works over silver-based samples, which resulted in better performance over $0.2Ag/Co_{0.93}Ce_{0.07}$ than over $0.2Ag/Co_3O_4$. Soot seemed to stimulate surface reduction at room temperature. The shift of Co $2p_{3/2}$ to a higher value also illustrates the reduction of cobalt with a higher oxidation state. Ce 3d spectrum provides similar clues. A characteristic peak for Ce^{3+} at 904 eV is observed over $0.2Ag/Co_{0.93}Ce_{0.07}$ and soot mixture [35, 41].

The durability tests were performed in NO_x (Fig. 4.14). Almost no loss of catalytic activity is detected after seven cycles, which illustrates its excellent reproducibility. The variation of pretreatment conditions provides a better understanding of the catalysts. The catalytic performance considerably decreases after purging with Helium at 700 °C for 1 h, and the ignition temperature rose to 260 °C. Interestingly, as the pretreated sample is cooled down to room temperature and exposed to the reactant gas for 3 h, the catalyst experiences a "regeneration" process. Namely, the catalytic performance recovers to the level of fresh materials. The analyses of FTIR and XPS provide more interpretation for the regeneration (Figs. 4.15 and 4.16). The peak at 1384 cm^{-1} which corresponds to nitrate species was observed in all the samples except the one pretreated at 700 °C in helium [42–45] and nitrite species disappear in the whole run [46, 47]. It is postulated that nitrate species served as a crucial factor in soot abatement in NO_x. XPS survey elucidates the oxidation state of silver. High-temperature treatment in Helium causes the symmetric Ag 3d peaks, which demonstrates the disappearance of $Ag^{\delta+}$. A high binding energy peak located at 369 eV (about 10%) appears over NO_x treated sample. This peak can be attributed to the formation of $AgNO_3$ [48, 49]. It is noteworthy that no $Ag^{\delta+}$ has been identified, probably because purging NO_x leads

Table 4.5 Statistics of XPS survey [19]

Catalysts	O_I (%)	O_{II} (%)	O_{III} (%)	A_o (%)①	Co $2p_{3/2}$ (eV)	$Ag^{\delta+}/Ag^*$ (%)②	Ag $3d_{5/2}$ (eV)
Co_3O_4	50.47	34.49	15.04	49.53	779.6	n/a	n/a
$Co_{0.93}Ce_{0.07}$	49.00	34.62	16.38	51.00	779.1	n/a	n/a
$0.2Ag/Co_3O_4$	43.87	44.40	11.72	56.13	779.6	13.12	368.1
$0.2Ag/Co_{0.93}Ce_{0.07}$	40.02	56.33	3.65	59.98	779.1	7.09	368.0
$0.2Ag/Co_3O_4$ + soot	33.98	57.79	8.23	66.02	779.7	8.26	368.1
$0.2Ag/Co_{0.93}Ce_{0.07}$ + soot	32.96	59.20	7.84	67.04	779.4	4.85	368.0

①A_o shows the percentage of adsorbed oxygen, and it equals to O_{II} + O_{III}
②$Ag^{\delta+}/Ag^*$ reveals the ratio surface oxidized silver to the total amount of silver

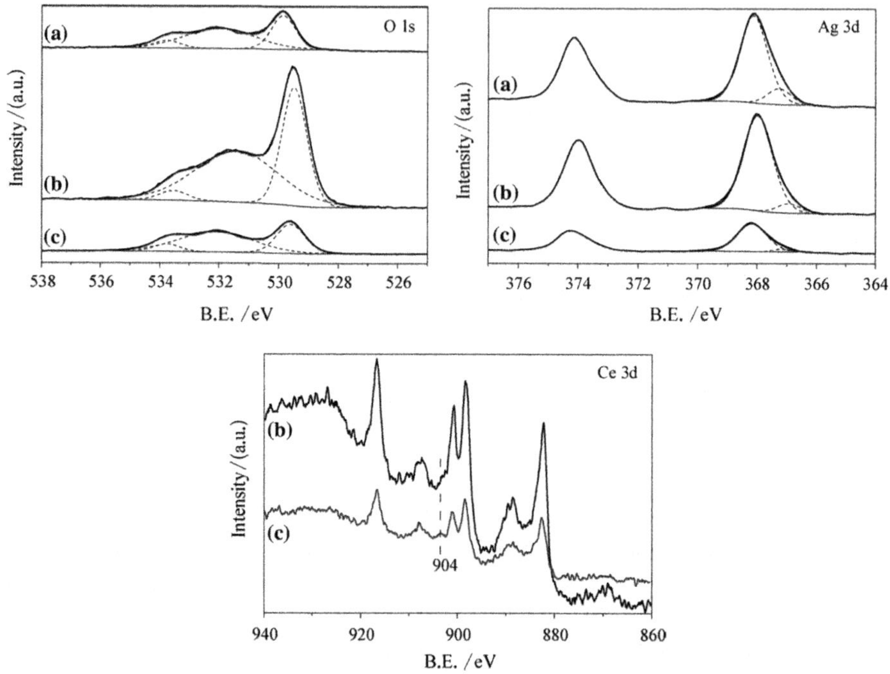

Fig. 4.13 XPS spectra of O 1 s, Ce 3d, and Ag 3d from Ag/Co–Ce complex oxides [19], **a** 0.2Ag/Co_3O_4; **b** 0.2Ag/$Co_{0.93}Ce_{0.07}$; **c** tight mixture of 0.2Ag/$Co_{0.93}Ce_{0.07}$ and Printex U soot

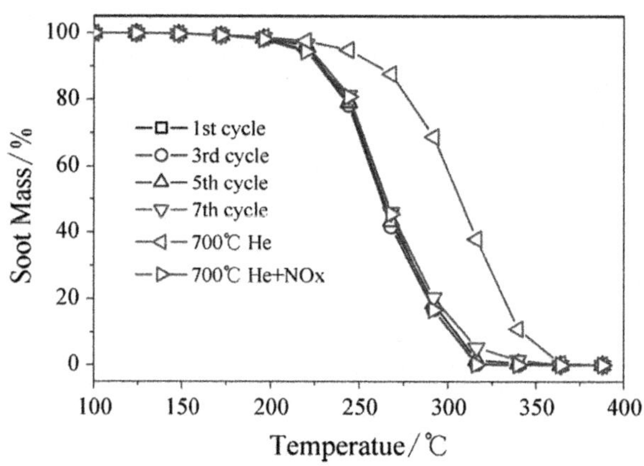

Fig. 4.14 Durability and stability tests for simultaneous removal of soot and NO_x over 0.2Ag/$Co_{0.93}Ce_{0.07}$ catalysts [19]. 700 °C He refers to fresh samples pretreated in helium at 700 °C, whereas 700 °C He + NO_x represents the previous sample exposure to reactant gas at room temperature for 3 h

to the conversion of silver oxides into $AgNO_3$. On the basis of catalytic performance and analysis of FTIR and XPS, $AgNO_3$ species are the key to soot gasification in NO_x.

The effect of silver can be categorized into the following three aspects, on the basis of literature reports and experimental results.

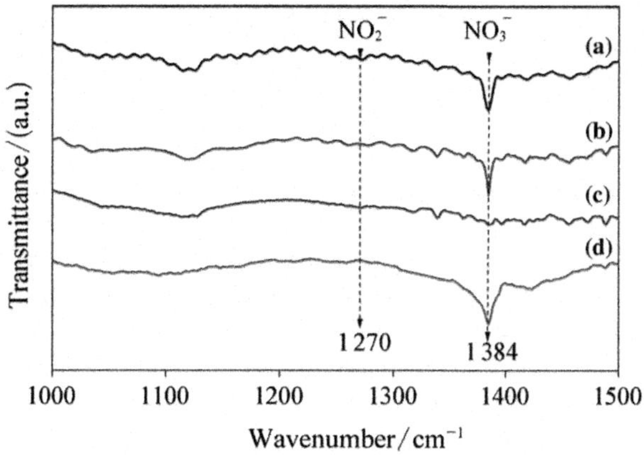

Fig. 4.15 FTIR spectra for $0.2Ag/Co_{0.93}Ce_{0.07}$ [19], **a** fresh sample, **b** after 7 cycles, **c** fresh sample pretreated by Helium for 1 h at 700 °C, **d** sample (**c**) pretreated with reactant gas (2000 ppm NO + 5%O_2) at room temperature for 3 h

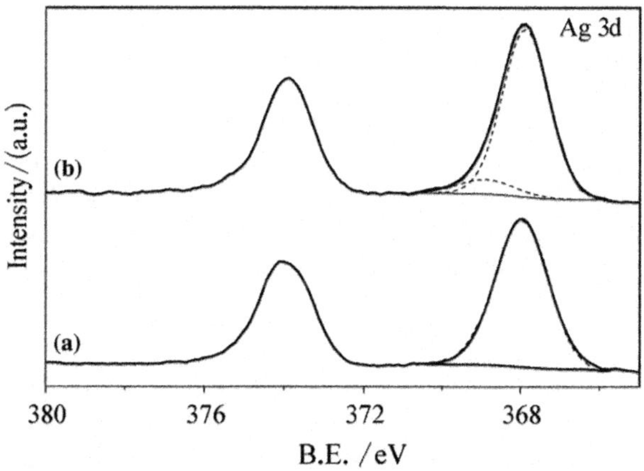

Fig. 4.16 XPS survey of $0.2Ag/Co_{0.93}Ce_{0.07}$ for Ag 3d spectra [19], **a** fresh sample pretreated by Helium for 1 h; **b** sample (**a**) pretreated by reactant gas (2000 ppm NO + 5%O_2) at room temperature

(1) Formation of oxidized silver in the surface.

Among a variety of the metal oxides, bulk Ag_2O is proved to be the most active one, yet the poor stability limits its application. The theoretical calculation also manifests that silver related oxygen species existed below 200 °C over the Ag(111) crystal plane [50]. Li and Yang [51] detect various oxidized silver species such as Ag_2O, Ag_2O_3, and Ag_3O over the crystal face of Ag(100) and Ag(111). The mixed silver oxides were highly active, and they were recognized as an essential factor in other catalytic oxidation reactions [52–54]. For soot removal, the effect of $Ag^{\delta+}$ has been elaborated by Haneda and Towata [55]. In the present study, the silver ions were even detected under the low partial pressure of O_2 in XPS survey (Fig. 4.16 and Table 4.5), which indicated the formation of surface silver oxides in the presence of O_2. This phenomenon also revealed the survival of silver oxides at room temperature and lower temperature. It can be summarized that the adsorbed oxygen on the surface can directly participate in the reaction over silver containing samples.

(2) Generation of active oxygen species.

Zero-valence silver is skilled at dissociating oxygen into active oxygen-containing peroxide and superoxide species [56, 57]. The combination of silver and various supports further promotes the amount of active oxygen. Superoxide species was discovered over Ag/activate carbon catalysts even in rich burn conditions [58]. The intensified superoxide signal for CeO_2 was observed with the assistance of silver [59]. In the present study, this process requires a relatively "clean" metallic silver surface. This "clean" surface is likely to be acquired at a higher temperature followed by the destruction of silver oxides species.

(3) Promotion of soot–catalyst contact conditions.

Silver nitrate is a low melting point compound; its melting point can reach as low as 210 °C, and its decomposition temperature is 440 °C [60]. The melting temperature of $AgNO_3$ is just close to the T_{10} value of the $0.2Ag/Co_{0.93}Ce_{0.07}$, which implies the probable correlation between the two values. Before the consumption of reducing agent, $AgNO_3$ in the mixture is in melting state, and $Ag^{\delta+}$ migration facilitates the soot–catalyst contact. This phenomenon is entirely different from various types of silver oxides. Ag_2O is decomposed after one run, while AgO is decomposed above 100 °C. Thus, silver oxides are unlikely to improve the contact conditions. However, the mobility of silver is to be taken into account. The mobility of silver particle is visualized by a variety of ETEM study, which is elaborated in Sect. 4.1.2.

Unlike most metal oxides materials, the xAg/$Co_{0.93}Ce_{0.07}$ catalyst performs even better in O_2 than in NO_x. The transitional oxidizing agents (silver oxides and silver nitrates) are keys to elucidating this result. The AgO_y stabilized in the presence of the O_2 atmosphere is quickly converted to $AgNO_3$ and/or other nitrate salts with the assistance of NO_x. The disappearance of AgO_y consequently goes against the catalytic performance. It can be inferred that some fast reactions for the catalytic process are missing in the absence of AgO_y; or rather, $AgNO_3$ is not as active as AgO_y. This deduction can be further confirmed according to the results shown in Fig. 4.17. The DTG signals uncover that Ag_2O, the primary phase of AgO_y, attacks soot at a lower temperature than $AgNO_3$ does in the O_2-rich and O_2-absent atmosphere. Ag_2O seems to be a stronger oxidant than $AgNO_3$. Hence, the untraditional catalytic behavior strongly depends on the activity of silver-based intermediates.

Highly active xAg/$Co_{0.93}Ce_{0.07}$ composite catalysts whose ignition temperature can be below 200 °C have been explored for catalytic soot combustion. Silver significantly enhanced catalytic behavior in various atmospheres. In O_2, the low-temperature redox cycle occurs between soot and the silver oxides situated on the surface of metallic silver. In the NO_x-rich atmosphere, these active silver oxides transform into $AgNO_3$ which is less active than silver oxides. The difference in

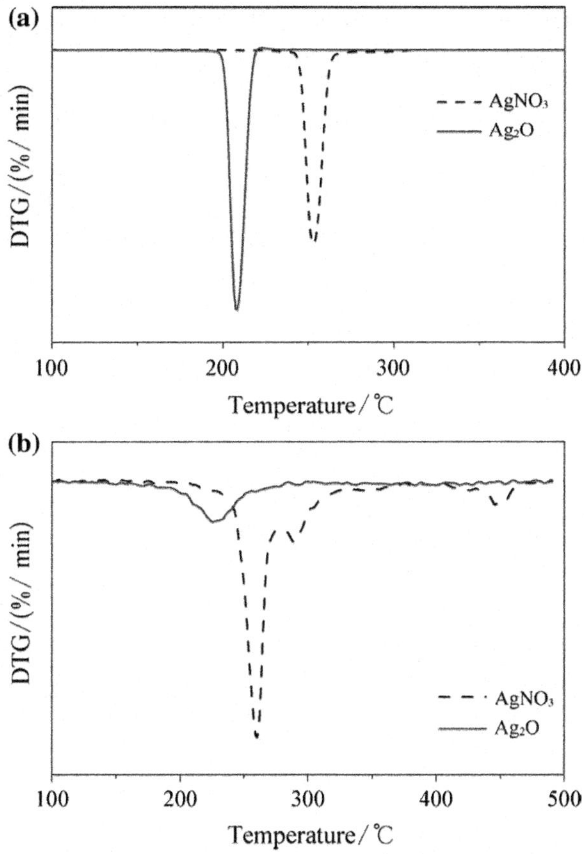

Fig. 4.17 DTG profiles for $AgNO_3$/soot mixture and Ag_2O/soot mixture in, **a** Ar and **b** 20% O_2 + 80% N_2 [19]

activity between the two transition state oxidization agent, Ag_2O, and $AgNO_3$ accounts for the untraditional catalytic performance in O_2 and NO_x over these catalysts.

4.3 Surface Area

The effect of surface area on the catalytic performance has been discovered since this approach was proposed. However, the address of the point is not on a deep-seated level, and the topic itself is still controversial. As mentioned above, some literature holds the view that the reaction takes place at three-phase boundary whereas other researchers believe the surface area is also a factor altering the catalytic behaviors.

Shimokawa and his co-workers investigated the effect of surface area on the catalytic performance of diesel soot gasification over La–K–Mn–O perovskite

Table 4.6 Preparation conditions, the surface area of $La_{0.8}K_{0.2}MnO_3$ [61]

Preparation method	Starting materials						
	La	K	Mn	Organic acids	Molar ratio of organic acids to metal cations	Calcination temperature (°C)	Specific surface area (S_a) ($m^2\ g^{-1}$)
(1)	Nitrate	Nitrate	Nitrate	Malic acid	1.5/1.0	550	18.5
						600	16.7
						750	14.7
(2)	Nitrate	Acetate	Acetate	n/a		850	8.2
						950	3.1
						1050	1.1

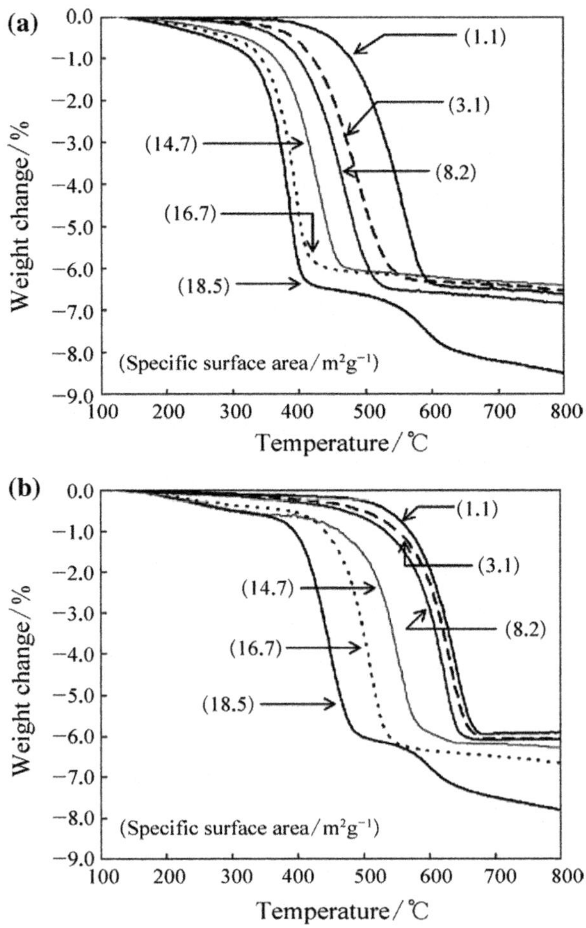

Fig. 4.18 TG curves of a mixture of carbon black (5 wt%) and $La_{0.8}K_{0.2}MnO_3$ with different surface areas under, **a** tight- and **b** loose contact conditions [61]

Table 4.7 Reaction results of $La_{0.8}K_{0.2}MnO_3$ in various contact modes [61]

Preparation method	Calcination temperature	Specific surface area (S_a) ($m^2\ g^{-1}$)	T_{max} (°C)	
			Tight contact	Loose contact
(1)	550	18.5	384	439
	600	16.7	396	503
	750	14.7	415	546
(2)	850	8.2	457	618
	950	3.1	485	628
	1050	1.1	535	638

catalysts [61]. The BET surface area is tuned via modifying the calcination temperature and precursors (Table 4.6). In this case, it is prominent that the surface area plays a decisive role in soot combustion (Fig. 4.18 and Table 4.7). The higher calcination temperature (above 850 °C) leads to the sintering of the material, thus worsening the catalytic performance.

Massive variables, including, preparation method, calcination temperature, and Ce/Zr ratios are tuned aiming at acquiring Ce–Zr with different BET surface areas; the catalytic behavior of these materials are also studied. The catalytic performance for soot removal in NO_x/O_2 is judged by both the BET surface area and the Ce/Zr atomic surface ratio of the catalysts, as shown in Fig. 4.19. As a general tendency, a relationship between $T_{50\%}$ and BET area, namely the higher the area, the lower the $T_{50\%}$, can be viewed. On the other hand, a wide dispersion of data points is observed, indicating that the BET area is not the sole factor affecting the catalytic performance. The Ce/Zr atomic ratio is also responsible for the overall performance. Three groups of Ce/Zr surface ratios have been identified, and two areas divided by the BET valve value (90 m^2/g) are obtained. For samples with BET surface area lower than 90 m^2/g, approximately, a reasonable linear relationship is drawn for each series of samples. This means that only when samples with similar Ce/Zr surface ratio are compared, the catalytic activity and the BET surface area follow a linear trend. On the other hand, if samples with close BET surface area are compared, those with a Ce/Zr surface ratio in the range of 2.0–3.1, which is just close to stoichiometric value (3.2), form an active group. The catalytic activity of materials with either a cerium-enriched surface (Ce/Zr surface ratio 3.9–7.6) or with a zirconium-enriched surface (Ce/Zr surface ratio 0.8–1.8) with regard to the nominal value exhibits lower activity than those with a more homogeneous distribution of the cations throughout the particles. These results demonstrate that the catalytic activity of soot combustion in NO_x/O_2 over $Ce_{0.76}Zr_{0.24}O_2$ mixed oxides depends both on the surface area and Ce/Zr surface properties. The surface distribution of cations is related to the homogeneity of the solid solution formed, which strongly depends on the preparation procedure. This is true for catalysts with BET surface area lower than 90 m^2/g approximately. However, constant values of the $T_{50\%}$ parameter are obtained for catalysts with BET surface areas above this

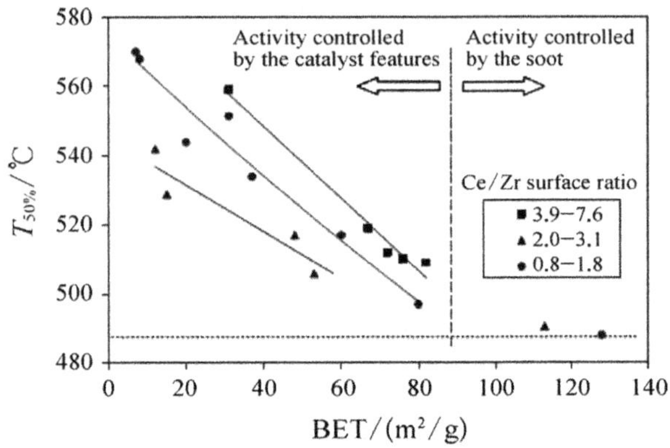

Fig. 4.19 $T_{50\%}$ parameter versus BET surface area of the catalysts [62]

threshold, that is, there is an achieved minimum temperature irrespective of BET value.

In another study, cerium-rich catalyst (CRX) and zirconium-rich catalyst (ZRX) were synthesized by co-precipitation method; their catalytic activity for soot combustion was also tested [63]. The relationship between Tm and surface area under different contact conditions are shown in Fig. 4.20. It is noticed that the influence of the surface area is less remarkable under loose contact condition than tight contact condition. Although there are more redox sites to create active oxygen on the high-surface-area samples, it is inaccessible for active oxygen to transfer from the catalyst to soot particles under loose contact conditions. On the other hand, a strong correlation between the catalytic activity of surface area is uncovered for the low-surface-area (<40 m^2/g) samples. As the surface area further expands, the catalytic activity is not allergic to the variation of surface area. It is also apparent that the catalytic activity of CRX is similar to that of ZRX under loose contact condition this situation is especially better under tight contact condition.

According to the above-presented studies, it is so challenging to elucidate the effect of surface area on the catalytic activity, since the surface area is not the sole variable. Many other parameters of the material are also coupled with the surface area. In most cases, additional one or two variables are introduced as the surface area is changed. These parameters as follows should be taken into account:

(1) The grain size of particles which is controlled by tuning the calcination temperature.
(2) The surface properties, for instance, oxygen vacancies and metal cations ratio, which are modified by preparation techniques.
(3) The exposure of active crystal face/efficient active site.

The study involving the effect of surface areas are profound and still in progress. According to the point of view, the research on the dependence of surface area against other parameters should be placed in the front rank.

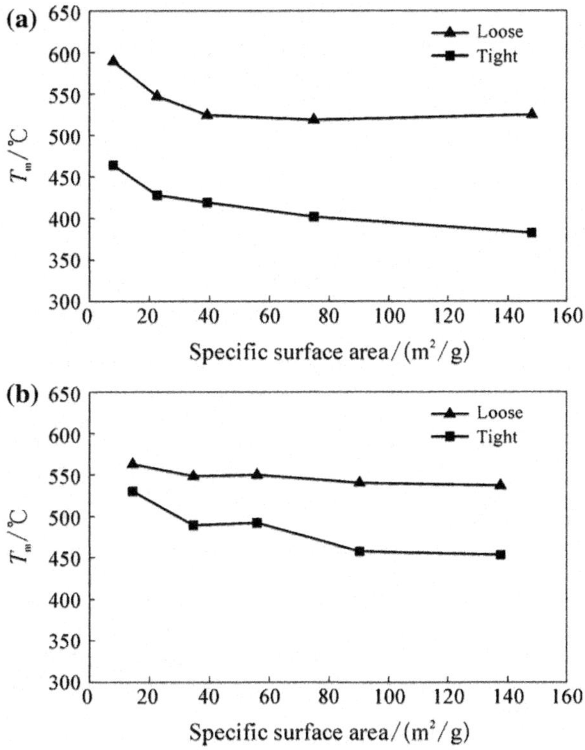

Fig. 4.20 Relationship between T_m and specific surface area over, **a** CRX and **b** ZRX series [63]

References

1. J. Van Doorn, J. Varloud, P. Mériaudeau, V. Perrichon, M. Chevrier, C. Gauthier, Effect of support material on the catalytic combustion of diesel soot particulates. Appl. Catal. B **1**, 117–127 (1992)
2. G. Mul, J.P. Neeft, F. Kapteijn, M. Makkee, J.A. Moulijn, Soot oxidation catalyzed by a Cu/K/Mo/Cl catalyst: evaluation of the chemistry and performance of the catalyst. Appl. Catal. B **6**, 339–352 (1995)
3. Y. Teraoka, S. Kagawa, Simultaneous catalytic removal of NO_x and diesel soot particulates. Catal. Surv. Asia **2**, 155–164 (1998)
4. J. Neeft, O.P. van Pruissen, M. Makkee, J.A. Moulijn, Catalysts for the oxidation of soot from diesel exhaust gases II. Contact between soot and catalyst under practical conditions. Appl. Catal. B **12**, 21–31 (1997)
5. E. Aneggi, V. Rico-Perez, C. de Leitenburg, S. Maschio, L. Soler, J. Llorca, A. Trovarelli, Ceria-zirconia particles wrapped in a 2D carbon envelope: improved low-temperature oxygen transfer and oxidation activity. Angew. Chem. **127**, 14246–14249 (2015)
6. L. Hensgen, K. Stöwe, Soot-catalyst contact studies in combustion processes using nano-scaled ceria as test material. Catal. Today **159**, 100–107 (2011)
7. S. Kureti, W. Weisweiler, K. Hizbullah, Simultaneous conversion of nitrogen oxides and soot into nitrogen and carbon dioxide over iron containing oxide catalysts in diesel exhaust gas. Appl. Catal. B **43**, 281–291 (2003)

8. S. Bensaid, N. Russo, D. Fino, CeO_2 catalysts with fibrous morphology for soot oxidation: the importance of the soot-catalyst contact conditions. Catal. Today **216**, 57–63 (2013)
9. G. Zhang, Z. Zhao, J. Xu, J. Zheng, J. Liu, G. Jiang, A. Duan, H. He, Comparative study on the preparation, characterization and catalytic performances of 3DOM Ce-based materials for the combustion of diesel soot. Appl. Catal. B **107**, 302–315 (2011)
10. Y. Wei, J. Liu, Z. Zhao, Y. Chen, C. Xu, A. Duan, G. Jiang, H. He, Highly active catalysts of gold nanoparticles supported on three-dimensionally ordered macroporous $LaFeO_3$ for soot oxidation. Angew. Chem. Int. Ed. **50**, 2326–2329 (2011)
11. G. Zhang, Z. Zhao, J. Liu, G. Jiang, A. Duan, J. Zheng, S. Chen, R. Zhou, Three dimensionally ordered macroporous $Ce_{1-x}Zr_xO_2$ solid solutions for diesel soot combustion. Chem. Commun. **46**, 457–459 (2010)
12. J. Xu, J. Liu, Z. Zhao, J. Zheng, G. Zhang, A. Duan, G. Jiang, Three-dimensionally ordered macroporous $LaCo_xFe_{1-x}O_3$ perovskite-type complex oxide catalysts for diesel soot combustion. Catal. Today **153**, 136–142 (2010)
13. S. Simonsen, S. Dahl, E. Johnson, S. Helveg, Ceria-catalyzed soot oxidation studied by environmental transmission electron microscopy. J. Catal. **255**, 1–5 (2008)
14. K. Nakamura, H. Oki, R. Sanui, N. Hidaka, M. Tanaka, H. Matsumoto, K. Hanamura, Characteristics of soot oxidation at the interface between soot and silicon-oxy-carbide with embedded single nanosized Pt particles. SAE Technical Paper (2013)
15. D. Gardini, J.M. Christensen, C.D. Damsgaard, A.D. Jensen, J.B. Wagner, Visualizing the mobility of silver during catalytic soot oxidation. Appl. Catal. B **183**, 28–36 (2016)
16. T. Epicier, A. Serve, M. Aouine, F.C.S. Aires, M. Tsampas, B. Cartoixa, K. Pajot, P. Vernoux, Environmental TEM investigation of the mechanism of soot combustion by Ag supported catalysts
17. E. Saab, E. Abi-Aad, M. Bokova, E. Zhilinskaya, A. Aboukaïs, EPR characterisation of carbon black in loose and tight contact with Al_2O_3 and CeO_2 catalysts. Carbon **45**, 561–567 (2007)
18. E. Kukovitsky, S. L'vov, N. Sainov, V. Shustov, L. Chernozatonskii, Correlation between metal catalyst particle size and carbon nanotube growth. Chem. Phys. Lett. **355**, 497–503 (2002)
19. G. Zou, Z. Fan, X. Yao, Y. Zhang, Z. Zhang, M. Chen, W. Shangguan, Catalytic performance of Ag/Co-Ce composite oxides during soot combustion in O_2 and NO_x: insights into the effects of silver (2017)
20. R. Matarrese, L. Castoldi, L. Lietti, P. Forzatti, Soot combustion: Reactivity of alkaline and alkaline earth metal oxides in full contact with soot. Catal. Today **136**, 11–17 (2008)
21. J. Oi-Uchisawa, S. Wang, T. Nanba, A. Ohi, A. Obuchi, Improvement of Pt catalyst for soot oxidation using mixed oxide as a support. Appl. Catal. B **44**, 207–215 (2003)
22. J. Oi-Uchisawa, A. Obuchi, S. Wang, T. Nanba, A. Ohi, Catalytic performance of Pt/MOx loaded over SiC-DPF for soot oxidation. Appl. Catal. B **43**, 117–129 (2003)
23. Y. Wei, J. Liu, Z. Zhao, C. Xu, A. Duan, G. Jiang, Structural and synergistic effects of three-dimensionally ordered macroporous $Ce_{0.8}Zr_{0.2}O_2$-supported Pt nanoparticles on the catalytic performance for soot combustion. Appl. Catal. A **453**, 250–261 (2013)
24. A. Setiabudi, B.A. Van Setten, M. Makkee, J.A. Moulijn, The influence of NO_x on soot oxidation rate: molten salt versus platinum. Appl. Catal. B **35**, 159–166 (2002)
25. E. Xue, K. Seshan, J. Ross, Roles of supports, Pt loading and Pt dispersion in the oxidation of NO to NO_2 and of SO_2 to SO_3. Appl. Catal. B **11**, 65–79 (1996)
26. S. Mulla, N. Chen, L. Cumaranatunge, G. Blau, D. Zemlyanov, W. Delgass, W. Epling, F. Ribeiro, Reaction of NO and O_2 to NO_2 on Pt: kinetics and catalyst deactivation. J. Catal. **241**, 389–399 (2006)
27. N. Güngör, S. Işçi, E. Günister, W. Miśta, H. Teterycz, R. Klimkiewicz, Characterization of sepiolite as a support of silver catalyst in soot combustion. Appl. Clay Sci. **32**, 291–296 (2006)
28. K.-I. Shimizu, H. Kawachi, A. Satsuma, Study of active sites and mechanism for soot oxidation by silver-loaded ceria catalyst. Appl. Catal. B **96**, 169–175 (2010)

29. E. Aneggi, J. Llorca, C. de Leitenburg, G. Dolcetti, A. Trovarelli, Soot combustion over silver-supported catalysts. Appl. Catal. B **91**, 489–498 (2009)
30. K. Yamazaki, T. Kayama, F. Dong, H. Shinjoh, A mechanistic study on soot oxidation over CeO_2–Ag catalyst with 'rice-ball' morphology. J. Catal. **282**, 289–298 (2011)
31. K.-I. Shimizu, M. Katagiri, S. Satokawa, A. Satsuma, Sintering-resistant and self-regenerative properties of Ag/SnO_2 catalyst for soot oxidation. Appl. Catal. B **108**, 39–46 (2011)
32. K. Villani, R. Brosius, J.A. Martens, Catalytic carbon oxidation over Ag/Al_2O_3. J. Catal. **236**, 172–175 (2005)
33. B. Dernaika, D. Uner, A simplified approach to determine the activation energies of uncatalyzed and catalyzed combustion of soot. Appl. Catal. B **40**, 219–229 (2003)
34. W. Shangguan, Y. Teraoka, S. Kagawa, Promotion effect of potassium on the catalytic property of $CuFe_2O_4$ for the simultaneous removal of NO_x and diesel soot particulate. Appl. Catal. B **16**, 149–154 (1998)
35. J. Liu, Z. Zhao, J. Wang, C. Xu, A. Duan, G. Jiang, Q. Yang, The highly active catalysts of nanometric CeO_2-supported cobalt oxides for soot combustion. Appl. Catal. B **84**, 185–195 (2008)
36. M. O'Connell, A. Norman, C. Hüttermann, M. Morris, Catalytic oxidation over lanthanum-transition metal perovskite materials. Catal. Today **47**, 123–132 (1999)
37. S. Shinde, K. Rajpure, X-ray photoelectron spectroscopic study of catalyst based zinc oxide thin films. J. Alloy. Compd. **509**, 4603–4607 (2011)
38. G. Zou, Y. Xu, S. Wang, M. Chen, W. Shangguan, The synergistic effect in Co–Ce oxides for catalytic oxidation of diesel soot. Catal. Sci. Technol. **5**, 1084–1092 (2015)
39. X. Bao, M. Muhler, T. Schedel-Niedrig, R. Schlögl, Interaction of oxygen with silver at high temperature and atmospheric pressure: A spectroscopic and structural analysis of a strongly bound surface species. Phys. Rev. B **54**, 2249 (1996)
40. G.B. Hoflund, Z.F. Hazos, G.N. Salaita, Surface characterization study of Ag, AgO, and Ag_2O using X-ray photoelectron spectroscopy and electron energy-loss spectroscopy. Phys. Rev. B **62**, 11126 (2000)
41. E. Bêche, P. Charvin, D. Perarnau, S. Abanades, G. Flamant, Ce 3d XPS investigation of cerium oxides and mixed cerium oxide ($Ce_xTi_yO_z$). Surf. Interface Anal. **40**, 264–267 (2008)
42. M. Grube, J. Lin, P. Lee, S. Kokorevicha, Evaluation of sewage sludge-based compost by FT-IR spectroscopy. Geoderma **130**, 324–333 (2006)
43. E. Smidt, K. Meissl, The applicability of Fourier transform infrared (FT-IR) spectroscopy in waste management. Waste Manag. **27**, 268–276 (2007)
44. V.G. Milt, E.D. Banús, M.A. Ulla, E.E. Miró, Soot combustion and NO_x adsorption on Co, Ba, K/ZrO_2. Catal. Today **133**, 435–440 (2008)
45. E. Smidt, P. Lechner, M. Schwanninger, G. Haberhauer, M. Gerzabek, Characterization of waste organic matter by FT-IR spectroscopy: application in waste science. Appl. Spectrosc. **56**, 1170–1175 (2002)
46. U. Bentrup, A. Brückner, M. Richter, R. Fricke, NO_x adsorption on MnO_2/NaY composite: an in situ FTIR and EPR study. Appl. Catal. B **32**, 229–241 (2001)
47. E. Ivanova, K. Hadjiivanov, D. Klissurski, M. Bevilacqua, T. Armaroli, G. Busca, FTIR study of species arising after NO adsorption and NO + O_2 co-adsorption on CoY: comparison with Co-ZSM-5. Micropor. Mesopor. Mater. **46**, 299–309 (2001)
48. H. Koga, T. Kitaoka, H. Wariishi, In situ synthesis of silver nanoparticles on zinc oxide whiskers incorporated in a paper matrix for antibacterial applications. J. Mater. Chem. **19**, 2135–2140 (2009)
49. T.-X. Liu, X.-Z. Li, F.-B. Li, $AgNO_3$-induced photocatalytic degradation of odorous methyl mercaptan in gaseous phase: mechanism of chemisorption and photocatalytic reaction. Environ. Sci. Technol. **42**, 4540–4545 (2008)
50. W.-X. Li, C. Stampfl, M. Scheffler, Why is a noble metal catalytically active? The role of the O–Ag interaction in the function of silver as an oxidation catalyst. Phys. Rev. Lett. **90**, 256102 (2003)

51. L. Li, J.C. Yang, Complex oxide structures formed by oxidation of Ag (100) and Ag (111) by hyperthermal atomic oxygen. Mater. High Temp. **20**, 601–606 (2003)
52. G. Lu, X. Zuo, Epoxidation of propylene by air over modified silver catalyst. Catal. Lett. **58**, 67–70 (1999)
53. R.E. Kenson, M. Lapkin, Kinetics and mechanism of ethylene oxidation. Reactions of ethylene and ethylene oxide on a silver catalyst. J. Phys. Chem. **74**, 1493–1502 (1970)
54. I.E. Wachs, R.J. Madix, The oxidation of methanol on a silver (110) catalyst. Surf. Sci. **76**, 531–558 (1978)
55. M. Haneda, A. Towata, Catalytic performance of supported Ag nano-particles prepared by liquid phase chemical reduction for soot oxidation. Catal. Today **242**, 351–356 (2015)
56. Y. Ono, T. Matsumura, N. Kitajima, S. Fukuzumi, Formation of superoxide ion during the decomposition of hydrogen peroxide on supported metals. J. Phys. Chem. **81**, 1307–1311 (1977)
57. R. Clarkson, A. Cirillo Jr., The formation and reactivity of oxygen as O_2^- on a supported silver surface. J. Catal. **33**, 392–401 (1974)
58. L. Chen, D. Ma, X. Bao, Hydrogen treatment-induced surface reconstruction: formation of superoxide species on activated carbon over Ag/activated carbon catalysts for selective oxidation of CO in H_2-rich gases. J. Phys. Chem. C **111**, 2229–2234 (2007)
59. M. Machida, Y. Murata, K. Kishikawa, D. Zhang, K. Ikeue, On the reasons for high activity of CeO_2 catalyst for soot oxidation. Chem. Mater. **20**, 4489–4494 (2008)
60. W. Davis, S. Rogers, A. Ubbelohde, Melting and crystal structure. The mechanism of melting of group I nitrates. Proc. R. Soc. Lond. Ser. A. Math. Phys. Sci. **220**, 14–24 (1953)
61. H. Shimokawa, H. Kusaba, H. Einaga, Y. Teraoka, Effect of surface area of La–K–Mn–O perovskite catalysts on diesel particulate oxidation. Catal. Today **139**, 8–14 (2008)
62. I. Atribak, A. Bueno-Lopez, A. Garcia-Garcia, Further insights into the key features of ceria-zirconia mixed oxides governing the catalysed soot combustion under NO_x/O_2. Top. Catal. **52**, 2088–2091 (2009)
63. Q. Liang, X. Wu, X. Wu, D. Weng, Role of surface area in oxygen storage capacity of ceria–zirconia as soot combustion catalyst. Catal. Lett. **119**, 265–270 (2007)

Chapter 5
Conclusion and Prospects

This book focuses on research on the simultaneous NO_x–soot removal via catalysis to purify the exhaust from diesel engines. Diesel engines have been widely popularized as a power source for vehicles owing to its reliable horsepower and excellent fuel economy. However, the emission from diesel engines is still a teaser. Simultaneous NO_x–soot removal via catalysis is one of the fundamental reactions, which serves as a bridge between soot and NO_x, one pair of pollutions with trade-off relations. Extensive concerns have risen to this strategy since current diesel engine after-treatment processes suffering severe problems, such as complicated system and byproduct disposal. To date, various catalysts have been investigated to achieve the above goal. The advancement of studies in this field promises possibilities of utilizing the one basket strategy, but at the same time also reveals challenges and difficulties that need to be overcome.

In Chap. 2, we summarized the performance of spinel structure material, perovskite-type material, hydrotalcite, rare-earth oxides, and mesoporous metal oxide as they were considered as potential candidates. Their surface properties ensure the provision of active sites in an oxidizing atmosphere. In parallel, their crystal structures are vulnerably modified by the incorporation of other metals such as alkali elements, which in turn affect the surface properties. Additionally, the ignition temperature and N_2 formation also can be tuned by this practice. It is a relatively ideal approach through which simultaneous catalytic removal of soot and NO_x can be achieved. A variety of promising materials were uncovered via the establishment of structure–activity relationship. However, it is still tricky for application in the actual diesel engine. Catalytic performances especially for their low-temperature activity are the primary challenge.

So how can we achieve our goal? The kinetic study in this reaction in Chap. 3 is significative to interpret the apparent activation energy of various materials, to elucidate the reaction order, and to further evidence the assumed mechanism. The book suggested that simultaneous NO_x–soot removal via catalysis to purify the exhaust from diesel engines could be classified into a typical heterogeneous reaction involving both gas–solid and solid–solid interaction. The reaction rates still

W. Shangguan et al., *Simultaneous Catalytic Removal of Diesel Soot and* NO_x,
Energy and Environment Research in China,
https://doi.org/10.1007/978-981-10-7266-6_5

could be depicted in the form of Arrhenius plot. On the basis of this tool, the detailed study indicates that the activity for soot combustion (CO_2) formation is strongly dependent on reactant gas; thus, the insight into the reaction is tough and complicated. The understanding of simultaneous NO_x–soot removal mechanism investigated by the state-of-the-art experimental methods, such as kinetics analysis and theoretical method, such as DFT calculation is further summarized in Chap. 3.

The catalytic reactions of simultaneous NO_x–soot removal take place at the three-phase boundary of a solid catalyst, a solid soot, and gaseous reactant(s). The contact between the diesel soot and catalyst is a critical factor for the simultaneous NO_x–soot removal performance. Chapter 4 highlighted that many factors, for instance, the contact mode between soot and catalyst will drastically affect the catalytic performance. Besides, the movement of the silver particle was traced during soot gasification. Furthermore, an untraditional phenomenon was uncovered over a PGM/mixed oxides-type catalyst in O_2 and $NO + O_2$ and a possible explanation is given. A thorough understanding of the mechanism behind the factors mentioned above would lay firm foundations to develop three-way or four-way catalysts which can be applied to diesel-powered vehicles. Further, this technique can be implemented in the purification of $PM_{2.5}$ and NO_x derived from stationary sources.

This book provides a perspective to interpret a solid–solid–gas heterogeneous process comprehensively. The methodology and the theoretical analysis are expected to inspire and assist readers to develop a proper catalyst aiming at a specific application.

Printed by Printforce, the Netherlands